Lecture Notes in Computer Science 11905

Founding Editors

Gerhard Goos
Karlsruhe Institute of Technology, Karlsruhe, Germany
Juris Hartmanis
Cornell University, Ithaca, NY, USA

More information about this series at http://www.springer.com/series/7412

Florian Knoll · Andreas Maier ·
Daniel Rueckert · Jong Chul Ye (Eds.)

Machine Learning for Medical Image Reconstruction

Second International Workshop, MLMIR 2019
Held in Conjunction with MICCAI 2019
Shenzhen, China, October 17, 2019
Proceedings

Springer

Editors
Florian Knoll 🆔
New York University
New York, NY, USA

Daniel Rueckert 🆔
Imperial College London
London, UK

Andreas Maier 🆔
University of Erlangen-Nuremberg
Erlangen, Germany

Jong Chul Ye 🆔
Korea Advanced Institute of Science
and Technology
Daejeon, South Korea

ISSN 0302-9743   ISSN 1611-3349   (electronic)
Lecture Notes in Computer Science
ISBN 978-3-030-33842-8   ISBN 978-3-030-33843-5   (eBook)
https://doi.org/10.1007/978-3-030-33843-5

LNCS Sublibrary: SL6 – Image Processing, Computer Vision, Pattern Recognition, and Graphics

This Springer imprint is published by the registered company Springer Nature Switzerland AG
The registered company address is: Gewerbestrasse 11, 6330 Cham, Switzerland

Preface

We are proud to present the proceedings for the Second Workshop on Machine Learning for Medical Image Reconstruction (MLMIR 2019) which was held on October 17, 2019, in Shenzhen, China, as part of the 22nd Medical Image Computing and Computer Assisted Intervention (MICCAI 2019) conference.

Image reconstruction is currently undergoing a paradigm shift that is driven by advances in machine learning (ML). Whereas traditionally transform-based or optimization-based methods have dominated methods of image reconstruction, ML has opened up an opportunity for new data-driven approaches which have demonstrated a number of advantages over traditional approaches. In particular, deep learning techniques have shown significant potential for image reconstruction and offer interesting new approaches. Finally, ML approaches also offer the possibility for application-specific image reconstruction, e.g. in motion-compensated cardiac or fetal imaging.

After the first successful workshop last year, we observed the further need for a scientific meeting addressing this emerging topic in image reconstruction. In particular, we are very proud that we were able to attract researchers from various modalities ranging from CT and MRI to Ultrasound and Molecular Imaging. It is great to see how the progress in methods, mathematics, and algorithms brings those often independently working communities closer together.

The aim of this year's workshop was to drive scientific discussion on advanced ML techniques for image acquisition and image reconstruction further, to identify new opportunities for applications as well as challenges in the evaluation and validation of ML based reconstruction approaches. We were fortunate that Markus Haltmeier (University of Innsbruck), Dong Liang (Shenzhen Institute of Advanced Technology), and Yong Long (Shanghai Jiao Tong University) gave fascinating keynote lectures that summarized the state of the art in this emerging field. Finally, we received 32 submissions and accepted 24 papers for inclusion in the workshop. The topics of the accepted papers cover the full range of medical image reconstruction problems, and deep learning dominates the machine learning approaches that are used to tackle the reconstruction problems.

September 2019

Florian Knoll
Andreas Maier
Daniel Rueckert
Jong Chul Ye

Organization

Workshop Organizers

Florian Knoll   New York University, USA
Andreas Maier   Friedrich-Alexander-University of Erlangen-Nuremberg,
                         Germany
Daniel Rueckert   Imperial College London, UK
Jong Chul Ye   Korean Institute of Science and Technology, South Korea

Scientific Program Committee

Simon Arridge   University College London, UK
Jose Caballero   Twitter, UK
Joseph Cheng   Stanford University, USA
Tolga Cukur   Bilkent University, Turkey
Bruno De Man   GE, USA
Enhao Gong   Stanford University, USA
Bernadette Hahn   University of Würzburg, Germany
Jo Hajnal   King's College London, UK
Markus Haltmeier   University of Innsbruck, Austria
Kerstin Hammernik   Technical University Graz, Austria
Andreas Hauptmann   University College London, UK
Mathews Jacob   University of Iowa, USA
Dong Liang   Chinese Academy of Sciences, China
Mariappan Nadar   Siemens Healthcare, USA
Ozan Öktem   Royal Institute of Technology, Sweden
Claudia Prieto   King's College London, UK
Jinyi Qi   UC Davis, USA
Essam Rashed   British University in Egypt, Egypt
Matthew Rosen   Havard University, USA
Michiel Shaap   HeartFlow, USA
Jo Schlemper   Imperial College London, UK
Nicole Seiberlich   Case Western University, USA
Ge Wang   Rensselaer Polytechnic Institute, USA
Shanshan Wang   Shenzhen Institute of Advanced Technology, China
Tobias Würfl   Friedrich-Alexander-University Erlangen-Nuremberg,
                         Germany
Guang Yang   Imperial College London, UK
Greg Zaharchuk   Stanford University, USA
Bo Zhu   Havard University, USA

Contents

Deep Learning for Computed Tomography

Deep Learning for General Image Reconstruction

Deep Learning for Magnetic Resonance Imaging

Recon-GLGAN: A Global-Local Context Based Generative Adversarial Network for MRI Reconstruction

Balamurali Murugesan[1,2](✉) [iD], S. Vijaya Raghavan[2], Kaushik Sarveswaran[2] [iD], Keerthi Ram[2], and Mohanasankar Sivaprakasam[1,2]

[1] Indian Institute of Technology Madras (IITM), Chennai, India
[2] Healthcare Technology Innovation Centre (HTIC), IITM, Chennai, India
balamurali@htic.iitm.ac.in

Abstract. Magnetic resonance imaging (MRI) is one of the best medical imaging modalities as it offers excellent spatial resolution and soft-tissue contrast. But, the usage of MRI is limited by its slow acquisition time, which makes it expensive and causes patient discomfort. In order to accelerate the acquisition, multiple deep learning networks have been proposed. Recently, Generative Adversarial Networks (GANs) have shown promising results in MRI reconstruction. The drawback with the proposed GAN based methods is it does not incorporate the prior information about the end goal which could help in better reconstruction. For instance, in the case of cardiac MRI, the physician would be interested in the heart region which is of diagnostic relevance while excluding the peripheral regions. In this work, we show that incorporating prior information about a region of interest in the model would offer better performance. Thereby, we propose a novel GAN based architecture, Reconstruction Global-Local GAN (Recon-GLGAN) for MRI reconstruction. The proposed model contains a generator and a context discriminator which incorporates global and local contextual information from images. Our model offers significant performance improvement over the baseline models. Our experiments show that the concept of a context discriminator can be extended to existing GAN based reconstruction models to offer better performance. We also demonstrate that the reconstructions from the proposed method give segmentation results similar to fully sampled images.

Keywords: Magnetic Resonance Imaging (MRI) · Reconstruction · Global local networks · Segmentation · Deep learning · Generative Adversarial Networks · Cardiac MRI

1 Introduction

Medical imaging is the preliminary step in many clinical scenarios. Magnetic resonance imaging (MRI) is one of the leading diagnostic modalities which can

---

Code available at https://github.com/Bala93/Recon-GLGAN.

© Springer Nature Switzerland AG 2019
F. Knoll et al. (Eds.): MLMIR 2019, LNCS 11905, pp. 3–15, 2019.
https://doi.org/10.1007/978-3-030-33843-5_1

produce images with excellent spatial resolution and soft tissue contrast. The major advantages of MRI include its non-invasive nature and the fact that it does not use radiation for imaging. However, the major drawback of MRI is the long acquisition time, which causes discomfort to patients and hinders applications in time critical diagnoses. This relatively slow acquisition process could result in significant artefacts due to patient movement and physiological motion. The slow acquisition time of MRI can be attributed to data samples not being collected directly in the image space but rather in k-space. k-space contains spatial-frequency information that is acquired line-by-line by the MRI hardware. In order to accelerate the MRI acquisition process, various methods ranging from Partial Fourier Imaging, Compressed Sensing and Dictionary Learning have been developed [4].

Recently, deep learning based methods have shown superior performance in many computer vision tasks. These methods have been successfully adapted for the MRI reconstruction problem and have shown promising results. The deep learning based methods [9] for MRI reconstruction can be broadly grouped into two: (1) k-space to image domain: the fully sampled image is obtained from zero-filled k-space. Examples include AUTOMAP and ADMM-Net. (2) image to image domain: the fully sampled (FS) image is obtained from the zero-filled (ZF) image. Our focus will be on the models of the latter kind. The work by Wang et al. [14] was the first to use convolutional neural networks to learn the mapping between ZF and FS images. Generative Adversarial Networks (GAN) [6] have shown promising results in many ill-posed inverse problems such as inpainting, super-resolution and denoising when compared to other deep learning based methods. The MRI reconstruction problem, having a similar problem formulation, has been approached with GANs and have shown encouraging results. The main focus of our paper is thus the application of GANs for the MRI reconstruction problem.

In the GANCS work [10], the generator is a residual network, the discriminator is a general deep network classifier and a combination of L1 and adversarial loss constitutes the loss function. Similarly, another work ReconGAN [11] uses a multi-stage network as a generator; a simple deep network classifier for the discriminator, and a combination of MSE loss in the image and frequency domains, adversarial loss constitute the loss function. The addition of the frequency domain loss adds data consistency. DAGAN [15] is another work which uses U-Net as a generator, a deep learning classifier as the discriminator with a combination of MSE loss in the image and frequency domains, adversarial loss and perceptual loss as the loss function. It showed that incorporating the perceptual loss term improved the reconstructed image quality in terms of the visually more convincing anatomical or pathological details. CDFNet [3] proposed the use of a combination of MSE loss in the image and frequency domains along with the Structural Similarity Index Measure (SSIM) as a loss function. This can be extended to a GAN setup. We will refer to this setup as ComGAN. SEGAN [8] proposed a generator network called SU-Net and used a general deep

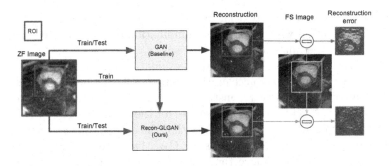

Fig. 1. Illustration depicting the comparison between the baseline GAN model and our Recon-GLGAN model. In the training phase, the ZF image and the ROI are fed in as inputs to the Recon-GLGAN model, while the baseline GAN only takes the ZF image as input. In the testing stage, the ZF image is fed as input to either model to produce the reconstruction (Note: ROI is not used during testing stage). The reconstruction error of the Recon-GLGAN model is lesser than the baseline GAN model in the ROI

network classifier as the discriminator. The loss term used is a combination of MSE in the image domain, SSIM and patch correlation regularization.

We refer to the concept of application-driven MRI as described in [2]: incorporating prior information about the end goal in the MRI reconstruction process would likely result in better performance. For instance, in the case of cardiac MRI reconstruction, the physician would be interested in the heart region, which is of diagnostic relevance while excluding the peripheral regions. Using this prior information about the region of interest (ROI) could lead to a better reconstruction. Another perspective is to note that the MRI reconstruction is not the goal in itself, but a means for further processing steps to extract relevant information such as segmentation or tissue characterisation. In general, segmentation algorithms would be interested in the specific ROI. Thus, incorporating prior information about the ROI in the reconstruction process would give two fold benefits: (1) The reconstruction would be better, (2) The segmentation algorithms consequently, could offer better results. The GAN based reconstruction methods described above did not incorporate the application perspective of MRI. Recently, [13] proposed a method in an application-driven MRI context, where the segmentation mask is obtained directly from a ZF image. This work showed encouraging results, but the model produces only the mask as output while the physician would be interested in viewing the FS image. Incorporating the ideas stated above, we propose a novel GAN based approach for MRI reconstruction. A brief outline of our approach compared to baseline GAN approaches is shown in Fig. 1. The key contributions of our work can be summarized as follows:

1. We propose a novel GAN architecture, Reconstruction Global-Local GAN (Recon-GLGAN) with a U-Net generator and a context discriminator. The context discriminator consists of a global feature extractor, local feature extractor and a classifier. The context discriminator architecture leverages

global as well as local contextual information from the image. We also propose a loss function which is a linear combination of context adversarial loss and L1 loss in the image domain.

2. We conducted extensive experiments to evaluate the proposed network with a context discriminator for acceleration factors of 2x, 4x and 8x. Our network showed significantly better reconstruction performance when compared with the baseline GAN and UNet architectures for the whole image as well as for a specific region of interest. We also show that the concept of a context discriminator can be easily extended to existing GAN based reconstruction architectures. To this end, we replace the discriminator in the existing GAN based reconstruction architectures with our context discriminator. This showed a significant performance improvement across metrics for an acceleration factor of 4x.

3. We conduct preliminary experiments to show that our model produces reconstructions that result in a better performance for the segmentation task. We demonstrate this using UNet model for segmentation, pre-trained on FS images and the corresponding masks. We observe that the segmentation results produced by the images from our Recon-GLGAN model are similar to FS images in comparison with the ZF and GAN images.

2   Methodology

2.1   Problem Formulation

Let $x_f \in C^N$ be the fully sampled complex image with dimensions $\sqrt{N} \times \sqrt{N}$ arranged in column-wise manner. x_f is obtained from fully sampled k-space measurements $(y_f \in C^N)$ through a fully sampled encoding matrix F_f using the relation $y_f = F_f x_f$. During undersampling, a subset of kspace measurements $(y_u \in C^M)$ say $(M \ll N)$ only are made. This corresponds to an undersampled image x_u by the relation $x_u = F_u^{-1} y_u$. x_u will be aliased due to sub-Nyquist sampling. Reconstructing x_f directly from y_u is ill-posed and direct inversion is not possible due to under-determined nature of system of equations. In our approach, we use deep learning network to learn the mapping between x_u and x_f. The neural network thus learns to minimize the error between predicted fully sampled image $(\hat{x}_f)$ and the ground truth (x_f).

2.2   Generative Adversarial Networks (GAN)

The GAN [6] consists of a generator (G) and discriminator (D). The generator (G) in GAN learns the mapping between two data distributions with the help of discriminator. In the case of MRI reconstruction, the goal of the generator is to learn the mapping between the data distribution of the ZF image (x_u) and FS image (x_f). The discriminator learns to distinguish between the generated and target reconstruction.

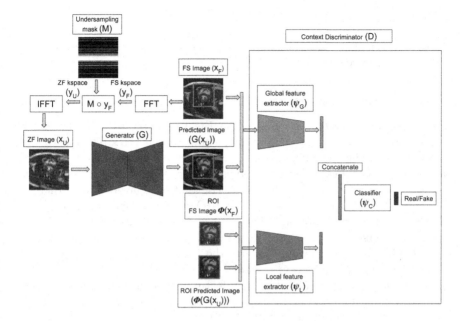

Fig. 2. Recon-GLGAN architecture

2.3   Proposed Reconstruction Global-Local GAN (Recon-GLGAN)

We propose a novel GAN architecture called Reconstruction Global-Local GAN (Recon-GLGAN). The idea is inspired from a GAN based work [5] in the context of image inpainting. The idea behind Recon-GLGAN is to capture both the global and local contextual features. Recon-GLGAN consists of a generator and a context discriminator. The generator (G) tries to learn the mapping between data distribution of ZF image x_u and FS image x_f with the help of the context discriminator which can extract global and local features and classify it as real/fake. The context discriminator consists of three components: global feature extractor, local feature extractor and classifier. The global feature extractor (Ψ_G) takes the entire image as input while the local feature extractor (Ψ_L) takes the region of interest (ROI) (Φ) from the entire image. The classifier network (Ψ_C) takes the concatenated feature vector ($\Psi_G(x)||\Psi_L(x)$) to classify the input image as real/fake. The overview of the proposed architecture is shown in Fig. 2. The joint optimization of the generator and context discriminator parameters is given by:

$$\min_{\theta_G} \max_{\theta_D} L_{Recon-GLGAN}(\theta_D, \theta_G) = E_{x_f \sim p_{\text{train}}(x_f)}[\log D_{\theta_D}(x_f)]$$

$$+ E_{x_u \sim p_G(x_u)}[-\log(D_{\theta_D}(G_{\theta_G}(x_u)))] \quad (1)$$

$$D_{\theta_D}(x) = \Psi_C(\Psi_G(x)||\Psi_L(\Phi(x))) \quad (2)$$

2.4 Network Architecture

Generator (G): The most commonly used encoder-decoder architecture U-Net [12] is used as the generator.

Context Discriminator (D)

- *Global feature extractor* (Ψ_G): The global feature extractor operates on the whole image. In our case, the input image dimension is 160×160. The stack of 3 convolutional layers followed by 2 fully connected layers is used as the global feature extractor. Leaky ReLu is used as an activation function for each layer. Average pooling is applied after each convolutional layer. The kernel size of convolutional layer is represented by: (Output channels, Input channels, height, width, stride, padding). The three convolution layers have the following parameters: (1) $(32, 1, 9, 9, 1, 0)$ (2) $(64, 32, 5, 5, 1, 0)$ (3) $(64, 64, 5, 5, 1, 0)$. The 2 fully connected layers converts the feature maps from convolutional layer to 64-dimensional feature vector.
- *Local feature extractor* (Ψ_L): The local feature extractor operates on the specific ROI of an image. In our case, the dimension of the ROI is 60×60. The architecture is largely similar to that of the global feature extractor except for the dimensions of the feature vector of the fully connected layer, which is modified according to the image dimensions. The output is a 64-dimensional feature vector.
- *Classifier* (Ψ_C): The outputs of the global and the local feature extractors are concatenated together into a single 128-dimensional vector, which is then passed to a single fully-connected layer, to output a single, continuous value. A sigmoid activation function is used so that this value is in the $[0, 1]$ range and represents the probability that the reconstruction is real/fake.

2.5 Loss Function

The loss function to accommodate our network design is given below:

$$L_{total} = \lambda_1 L_{imag} + \lambda_2 L_{context} \tag{3}$$

$$L_{imag} = E_{x_u, x_f}[\|x_f - G(x_u)\|_1] \tag{4}$$

$$L_{context} = E_{x_f}[log(D(x_f))] + E_{x_u}[-log(D(G(x_u)))] \tag{5}$$

where L_{imag} is the L1 loss between predicted and target fully sampled image, $L_{context}$ is the context adversarial loss.

3 Experiments and Results

3.1 Dataset

Automated Cardiac Diagnosis Challenge (ACDC) [1] is a cardiac MRI segmentation dataset. The dataset has 150 and 50 patient records for training and testing respectively. From the patient records, 2D slice images are extracted and cropped to 160×160. The extracted 2D slices amount to 1841 for training and 1076 for testing. The slices are normalized to the range (0–1). In the context of MRI reconstruction, the slice images are considered as FS images while the ZF images are obtained through cartesian undersampling masks corresponding to 2x, 4x and 8x accelerations.

The MR images in training set have their corresponding segmentation masks whereas the segmentation masks for MR images in test set are not publicly available. The dimensions of the ROI is set to 60×60 based on a study of the sizes of the segmentation masks in the training set. In the training phase, the center of the ROI for each slice is the midpoint of the closest bounding box of the corresponding segmentation mask.

3.2   Evaluation Metrics

Peak Signal-to-Noise Ratio (PSNR), Structural Similarity Index (SSIM), Normalised Mean Square Error (NMSE) metrics are used to evaluate the reconstruction quality for the entire image and its ROI. The segmentation quality is evaluated using Dice similarity coefficient (DICE) and Hausdorff distance (HD).

3.3   Implementation Details

The models were implemented in PyTorch. All models were trained for 150 epochs on two Nvidia GTX-1070 GPUs. Adam optimizer was used for the generator, with a learning rate of 0.001. Stochastic Gradient Descent optimizer was used for the discriminator, with a learning rate of $5e^{-3}$. For the loss term, $\lambda_2 = 4e^{-4}$, and $\lambda_1 = 1$.

The ROI for the MR images in the test set is obtained by following the algorithm described in [7]. This ROI information is not used for inference, it is used only to evaluate the ROI's reconstruction quality.

3.4   Results and Discussion

Reconstruction. To evaluate the proposed network, we perform the following experiments:

(1) We compare our proposed Recon-GLGAN with the baseline architecture GAN, U-Net, and the ZF images. The metrics for each model for the whole image as well as ROI are shown in Table 1. The results show that our model Recon-GLGAN performs better than the baseline GAN and U-Net across all metrics for all acceleration factors. We also note that our model offers appreciable performance improvement for 4x and 8x acceleration factors compared to 2x. This can be attributed to the fact that the image degradation in the case of 2x is not severe when compared with 4x and 8x. The qualitative comparison of ZF, GAN and Recon-GLGAN for different acceleration factors are shown in Fig. 3. In the Figure, it can be observed that reconstruction error of Recon-GLGAN

Table 1. Comparison of Recon-GLGAN with baseline architectures for 2x, 4x and 8x accelerations (FI-Full image)

			NMSE	PSNR	SSIM
2x	FI	Zero-filled	0.01997 ± 0.01	26.59 ± 3.19	0.8332 ± 0.06
		UNet	0.00959 ± 0.00	29.7 ± 2.97	0.9069 ± 0.03
		GAN	0.00958 ± 0.01	29.72 ± 3.03	0.9083 ± 0.03
		Recon-GLGAN	$\mathbf{0.00956 \pm 0.00}$	$\mathbf{29.74 \pm 3.0}$	$\mathbf{0.9108 \pm 0.03}$
	ROI	Zero-filled	0.01949 ± 0.02	25.48 ± 3.73	0.859 ± 0.05
		UNet	0.00952 ± 0.01	28.48 ± 3.03	0.9036 ± 0.04
		GAN	$\mathbf{0.00942 \pm 0.00}$	28.53 ± 3.12	0.904 ± 0.04
		Recon-GLGAN	0.00944 ± 0.01	$\mathbf{28.54 \pm 3.19}$	$\mathbf{0.9065 \pm 0.04}$
4x	FI	Zero-filled	0.03989 ± 0.03	23.65 ± 3.38	0.7327 ± 0.08
		UNet	0.01962 ± 0.01	26.62 ± 3.209	0.8419 ± 0.05
		GAN	0.01934 ± 0.01	26.68 ± 3.08	0.8465 ± 0.05
		Recon-GLGAN	$\mathbf{0.01905 \pm 0.01}$	$\mathbf{26.8 \pm 3.25}$	$\mathbf{0.8497 \pm 0.05}$
	ROI	Zero-filled	0.03886 ± 0.04	22.63 ± 3.87	0.7514 ± 0.07
		UNet	0.01931 ± 0.01	25.46 ± 3.35	0.8242 ± 0.06
		GAN	0.01925 ± 0.02	25.52 ± 3.38	0.8301 ± 0.06
		Recon-GLGAN	$\mathbf{0.01878 \pm 0.02}$	$\mathbf{25.66 \pm 3.26}$	$\mathbf{0.8327 \pm 0.06}$
8x	FI	Zero-filled	0.08296 ± 0.06	20.46 ± 3.24	0.6443 ± 0.09
		UNet	0.03353 ± 0.02	24.26 ± 2.71	0.7547 ± 0.07
		GAN	0.03359 ± 0.02	24.25 ± 2.71	0.7557 ± 0.07
		Recon-GLGAN	$\mathbf{0.03286 \pm 0.02}$	$\mathbf{24.32 \pm 2.68}$	$\mathbf{0.7562 \pm 0.07}$
	ROI	Zero-filled	0.07943 ± 0.08	19.47 ± 3.82	0.6435 ± 0.07
		UNet	0.03147 ± 0.02	23.31 ± 2.88	0.72 ± 0.07
		GAN	0.03129 ± 0.02	23.33 ± 2.92	$\mathbf{0.7294 \pm 0.07}$
		Recon-GLGAN	$\mathbf{0.03102 \pm 0.02}$	$\mathbf{23.34 \pm 2.82}$	0.7293 ± 0.07

for entire image and its ROI is better than GAN. But, it is evident that, the reconstruction error of Recon-GLGAN is significantly better than GAN in the ROI compared with the entire image. This behaviour can be attributed to the design of context discriminator which has a separate feature extraction path for specified ROI. The design of context discriminator enables the generator to specifically learn the ROI along with the entire image during the training phase.

(2) We attempt to show that the concept of a context discriminator can be extended to existing GAN based works for MRI reconstruction. The different GAN based architectures and their corresponding loss functions can be found in Table 2. In this experiment to ensure a fair comparison, the generator is set to U-Net, discriminator is set to global feature extractor (Ψ_G) followed by a classifier (Ψ_C) (basic discriminator) and the loss functions are taken from their respective works [3,8,10,15]. This arrangement means that the difference between the

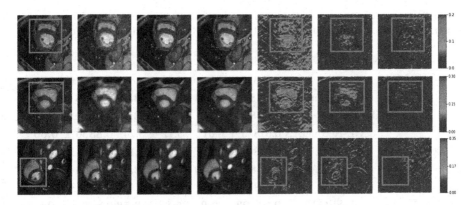

Fig. 3. From Left to Right: Ground Truth FS image, ZF image, GAN reconstructed image, Recon-GLGAN reconstructed image, ZF reconstruction error, GAN reconstruction error and Recon-GLGAN reconstruction error. From Top to Bottom: Images corresponding to different acceleration factors: 2x, 4x and 8x.

Table 2. GAN based reconstruction architectures and their loss terms

Architecture		Loss function terms
ReconGAN	–	L_{imag}, L_{global}, L_{freq}
	GL-ReconGAN	L_{imag}, $L_{context}$, L_{freq}
DAGAN	–	L_{imag}, L_{global}, L_{freq}, L_{vgg}
	GL-DAGAN	L_{imag}, $L_{context}$, L_{freq}, L_{vgg}
SEGAN	–	L_{imag}, L_{global}, L_{ssim}
	GL-SEGAN	L_{imag}, $L_{context}$, L_{ssim}
COMGAN	–	L_{imag}, L_{freq}, L_{global}, L_{ssim}
	GL-COMGAN	L_{imag}, L_{freq}, $L_{context}$, L_{ssim}

various GAN based architectures comes only from the generator loss. In this experiment, we replace the basic discriminator of the GAN architectures with our proposed context discriminator. The results comparing the GAN architectures with basic discriminator and context discriminator are reported in Table 3. From the Table, it is clear that the GAN with context discriminator have shown improved results compared to GAN with basic discriminator for different generator loss. A few sample results comparing the GAN based reconstruction methods with basic and context discriminator are shown in Fig. 4. From the figure we observe that the ROI's reconstruction error for GAN with context discriminator is lesser compared to GAN with the basic discriminator. This shows that the context discriminator can be extended to other GAN based reconstruction methods.

Table 3. Reconstruction metric comparison for full image and region of interest for various GAN based reconstruction architecture for 4x accelerations (FI - Full Image)

			NMSE	PSNR	SSIM
ReconGAN	FI	–	0.01857 ± 0.01	26.82 ± 2.89	0.8485 ± 0.05
		GL-ReconGAN	**0.01844 ± 0.01**	**26.91 ± 3.12**	**0.8498 ± 0.05**
	ROI	–	**0.018 ± 0.01**	**25.76 ± 3.06**	0.832 ± 0.06
		GL-ReconGAN	0.01836 ± 0.01	25.72 ± 3.24	**0.8336 ± 0.06**
SEGAN	FI	–	0.01862 ± 0.01	26.84 ± 3.10	0.8483 ± 0.06
		GL-SEGAN	**0.01817 ± 0.01**	**27.02 ± 3.4**	**0.8545 ± 0.05**
	ROI	–	0.0185 ± 0.01	25.64 ± 3.19	0.8308 ± 0.07
		GL-SEGAN	**0.01793 ± 0.01**	**25.87 ± 3.56**	**0.838 ± 0.06**
ComGAN	FI	–	0.01899 ± 0.01	26.78 ± 3.14	0.8481 ± 0.05
		GL-ComGAN	**0.01789 ± 0.01**	**27.06 ± 3.26**	**0.8505 ± 0.05**
	ROI	–	0.01872 ± 0.01	25.64 ± 3.28	0.8315 ± 0.06
		GL-ComGAN	**0.01766 ± 0.02**	**25.91 ± 3.25**	**0.834 ± 0.06**
DAGAN	FI	–	0.01903 ± 0.01	26.75 ± 3.06	0.8452 ± 0.06
		GL-DAGAN	**0.01851 ± 0.01**	**26.87 ± 3.03**	**0.845 ± 0.06**
	ROI	–	**0.01838 ± 0.01**	**25.68 ± 3.04**	0.8272 ± 0.07
		GL-DAGAN	0.01858 ± 0.01	25.62 ± 3.016	**0.8277 ± 0.07**

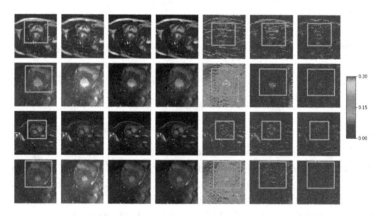

Fig. 4. From Left to Right: Ground Truth FS image, ZF image for 4x undersampling factor, GAN with basic discriminator reconstructed image, GAN with context discriminator reconstructed image, ZF reconstruction error, GAN with basic discriminator reconstruction error and GAN with context discriminator reconstruction error. From top to bottom: ReconGAN, SEGAN, ComGAN, DAGAN.

Segmentation. Image segmentation is an important task in medical imaging and diagnosis. For instance, in the case of cardiac MRI, the segmentation of left ventricle (LV), right ventricle (RV) and myocardium (MC) are used for cardiac

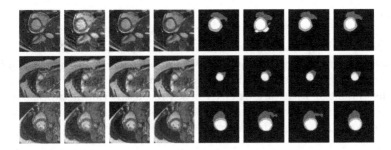

Fig. 5. From Left to Right: FS image, ZF image, GAN reconstructed image, Recon-GLGAN reconstructed image, Ground Truth FS segmentation mask, Segmentation mask for ZF, Segmentation mask for GAN reconstructed image and Segmentation mask for Recon-GLGAN reconstructed image. From top to bottom: Sample 1, 2 and 3

function analysis. Advances in deep learning networks have produced state-of-the-art results. These networks are trained on the FS images and, testing the network with ZF images will result in an unsatisfactory segmentation. We note that a better reconstruction, which is close to the FS image would result in better segmentation performance. In this experiment, we would like to show that the segmentation performance on the reconstructed images from our Recon-GLGAN model is better than the baseline GAN model. To demonstrate this, we use the most widely used segmentation network U-Net [12]. U-Net is trained on the FS images to produce multi-class (LV, RV and MC) segmentation outputs. Since the ground truth segmentation masks are unavailable for the test set of the ACDC dataset, we instead use the outputs of the FS images in the test set as ground truth. The reconstructed images from GAN and Recon-GLGAN are passed to the UNet and the corresponding segmentation masks are obtained. The obtained segmentation masks for sample images are shown in Fig. 5. It is evident from the figure that our network's performance is closest to FS followed by GAN and ZF images. The same are quantified using the segmentation metrics Dice and Hausdorff for the sample images in Fig. 6.

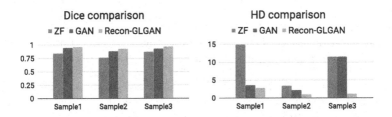

Fig. 6. Segmentation metrics: Dice and HD comparison for image samples 1, 2 and 3

4   Conclusion

In this work, we proposed a novel GAN network, Recon-GLGAN. The context discriminator proposed in Recon-GLGAN helps to capture both global and local features enabling a better overall reconstruction. We showed the extensibility of our discriminator with various GAN based reconstruction networks. We also demonstrated that the images obtained from our method gave segmentation results close to fully sampled images.

References

1. Bernard, O., Lalande, A., Zotti, C., Cervenansky, F., et al.: Deep learning techniques for automatic MRI cardiac multi-structures segmentation and diagnosis: is the problem solved? IEEE Trans. Med. Imaging **37**(11), 2514–2525 (2018)
2. Caballero, J., Bai, W., Price, A.N., Rueckert, D., Hajnal, J.V.: Application-driven MRI: joint reconstruction and segmentation from undersampled MRI data. In: Golland, P., Hata, N., Barillot, C., Hornegger, J., Howe, R. (eds.) MICCAI 2014. LNCS, vol. 8673, pp. 106–113. Springer, Cham (2014). https://doi.org/10.1007/978-3-319-10404-1_14
3. Dedmari, M.A., Conjeti, S., Estrada, S., Ehses, P., Stöcker, T., Reuter, M.: Complex fully convolutional neural networks for MR image reconstruction. In: Knoll, F., Maier, A., Rueckert, D. (eds.) MLMIR 2018. LNCS, vol. 11074, pp. 30–38. Springer, Cham (2018). https://doi.org/10.1007/978-3-030-00129-2_4
4. Hollingsworth, K.G.: Reducing acquisition time in clinical MRI by data undersampling and compressed sensing reconstruction. Phys. Med. Biol. **60**(21), R297–R322 (2015)
5. Iizuka, S., Simo-Serra, E., Ishikawa, H.: Globally and locally consistent image completion. ACM Trans. Graph. **36**(4), 107:1–107:14 (2017)
6. Isola, P., Zhu, J., Zhou, T., Efros, A.A.: Image-to-image translation with conditional adversarial networks. In: IEEE Conference on Computer Vision and Pattern Recognition (CVPR), pp. 5967–5976, July 2017
7. Khened, M., Kollerathu, V.A., Krishnamurthi, G.: Fully convolutional multi-scale residual densenets for cardiac segmentation and automated cardiac diagnosis using ensemble of classifiers. Med. Image Anal. **51**, 21–45 (2019)
8. Li, Z., Zhang, T., Zhang, D.: SEGAN: structure-enhanced generative adversarial network for compressed sensing MRI reconstruction. CoRR abs/1902.06455 (2019)
9. Lundervold, A.S., Lundervold, A.: An overview of deep learning in medical imaging focusing on MRI. Zeitschrift für Medizinische Physik **29**(2), 102–127 (2019)
10. Mardani, M., et al.: Deep generative adversarial neural networks for compressive sensing MRI. IEEE Trans. Med. Imaging **38**(1), 167–179 (2019)
11. Quan, T.M., Nguyen-Duc, T., Jeong, W.: Compressed sensing MRI reconstruction using a generative adversarial network with a cyclic loss. IEEE Trans. Med. Imaging **37**(6), 1488–1497 (2018)
12. Ronneberger, O., Fischer, P., Brox, T.: U-Net: convolutional networks for biomedical image segmentation. In: Navab, N., Hornegger, J., Wells, W.M., Frangi, A.F. (eds.) MICCAI 2015. LNCS, vol. 9351, pp. 234–241. Springer, Cham (2015). https://doi.org/10.1007/978-3-319-24574-4_28

13. Schlemper, J., et al.: Cardiac MR segmentation from undersampled k-space using deep latent representation learning. In: Frangi, A.F., Schnabel, J.A., Davatzikos, C., Alberola-López, C., Fichtinger, G. (eds.) MICCAI 2018. LNCS, vol. 11070, pp. 259–267. Springer, Cham (2018). https://doi.org/10.1007/978-3-030-00928-1_30

14. Wang, S., et al.: Accelerating magnetic resonance imaging via deep learning. In: IEEE 13th International Symposium on Biomedical Imaging (ISBI), pp. 514–517, April 2016

15. Yang, G., et al.: DAGAN: deep de-aliasing generative adversarial networks for fast compressed sensing MRI reconstruction. IEEE Trans. Med. Imaging **37**(6), 1310–1321 (2018)

Self-supervised Recurrent Neural Network for 4D Abdominal and In-utero MR Imaging

Tong Zhang[1]($\boxtimes$), Laurence H. Jackson[1], Alena Uus[1], James R. Clough[1], Lisa Story[2], Mary A. Rutherford[1], Joseph V. Hajnal[1], and Maria Deprez[1]

[1] School of Biomedical Engineering and Imaging Sciences, King's College London, London, UK
tong.zhang@kcl.ac.uk
[2] Department of Women and Children's Health, School of Life Course Sciences, King's College London, London, UK

Abstract. Accurately estimating and correcting the motion artifacts are crucial for 3D image reconstruction of the abdominal and in-utero magnetic resonance imaging (MRI). The state-of-art methods are based on slice-to-volume registration (SVR) where multiple 2D image stacks are acquired in three orthogonal orientations. In this work, we present a novel reconstruction pipeline that only needs one orientation of 2D MRI scans and can reconstruct the full high-resolution image without masking or registration steps. The framework consists of two main stages: the respiratory motion estimation using a self-supervised recurrent neural network, which learns the respiratory signals that are naturally embedded in the asymmetry relationship of the neighborhood slices and cluster them according to a respiratory state. Then, we train a 3D deconvolutional network for super-resolution (SR) reconstruction of the sparsely selected 2D images using integrated reconstruction and total variation loss. We evaluate the classification accuracy on 5 simulated images and compare our results with the SVR method in adult abdominal and in-utero MRI scans. The results show that the proposed pipeline can accurately estimate the respiratory state and reconstruct 4D SR volumes with better or similar performance to the 3D SVR pipeline with less than 20% sparsely selected slices. The method has great potential to transform the 4D abdominal and in-utero MRI in clinical practice.

1 Introduction

Due to its non-invasive nature and the superior soft-tissue contrast, magnetic resonance imaging (MRI) is becoming increasingly more popular for the adjunct pregnancy screening [1–4]. The typical scanning time for a 2D MR stack that covers the whole fetus and placenta varies from 1 to 10 min per stack depending on the MR sequences, field-of-view (FoV) and slice thickness. The long scanning time inevitably introduces a series of motion artifacts, such as maternal breathing, organ deformation, and fetal movement.

© Springer Nature Switzerland AG 2019
F. Knoll et al. (Eds.): MLMIR 2019, LNCS 11905, pp. 16–24, 2019.
https://doi.org/10.1007/978-3-030-33843-5_2

In Utero MR Image Reconstruction. To reconstruct a high quality 3D or 4D placenta and in-utero MRI, accurately estimating the respiratory motion is a key step. The current state-of-the-art methods correct the through-plane motion using slice-to-volume registration (SVR) reconstruction pipelines [5–9]. These methods require three to nine 2D image stacks to be acquired in orthogonal orientations; then a region-of-interest (ROI) mask needs to be generated manually or automatically [8,9]; finally the 3D SR image is iteratively generated based on the optimisation of the SVR results, robust statistics, intensity correction, and estimated point spread function (PSF). The acquisition of multiple 2D MR stacks in different orientations is time-consuming compared to the single orientation and inevitably introduces motion artifacts. The registration methods in the SVR pipelines are rigid and cannot correct the non-rigid respiratory motion. The SVR pipelines thus need enough redundant 2D images to reject all the slices where deformation compared to the reconstructed volume occurred. The overall image reconstruction performance will depend on the accuracy of the ROI masking, registration and data redundancy.

Self-supervised learning is generally considered as a subset of unsupervised learning, where the extensive cost of manual annotations is avoided and replaced by supervisory signals or automatically generated labels. Compared to the popular supervised methods which train the neural network with paired data Xi and label Yi, the self-supervised methods train with data Xi with its pseudo label Pi, which is generated automatically without involving any human annotation. Several recent papers have explored the usage of the temporal ordering of frames/images as a supervisory signal for complex video analysis [10–12]. In particular, Wei et al. [12] explored detecting and learning the direction of time for action recognition and video forensics.

Contribution. In this work, we propose a respiratory motion resolved 4D (3D+t) reconstruction pipeline of a single orientation stack of 2D MR slices, based on an bidirectional self-supervised recurrent neural network (RNN) [13] for identification of breathing states and efficient modified balanced steady state free precession (bSSFP) sequence with the SWEEP technique [14]. The method does not require masking or registration. Our experimental results show that the SWEEP MR acquisition in combination with the proposed pipeline enables 4D (3D+t) SR reconstruction of abdominal and in-utero images, and outperforms the SVR for 3D reconstruction with using less than 20% total slices for each respiratory state. To the best of our knowledge, it is the first successful application of self-supervised network for image-driven respiratory motion estimation and 4D(3D+t) MR SR reconstruction in the medical imaging community.

2 Method

2.1 Data Acquisition

A stack of MR slices is acquired sequentially using a modified bSSFP SWEEP sequence [14] which allows fast acquisition of large number of densely spaced overlapping slices, thus providing sufficient information for local estimation of respiratory motion. SWEEP continuously shifts the radiofrequency excitation frequency so as to maintain a single stable signal state across a volume, negating the requirement for start-up cycles and resulting in a maximally efficient acquisition for dense slice sampling applications. The acquisition time per slice is 490 ms for the uterus scans and 442 ms for the kidney scans, which freezes nearly all in-plane respiratory motion. The total scan time depends on the total slice number which is 3 to 10 min. This sequence also minimises the effects of fetal motion, by minimising the time between acquisition of the neighbouring slices while maintaining high MR signal. This effectively removes the need for masking, as the data is locally consistent except for the respiratory motion.

2.2   The Reconstruction Pipeline

The reconstruction pipeline consists of cascading a self-supervised RNN to estimate the respiratory states for each slice and a three layer super-resolution (SR) neural network (SR-net) for reconstruction respiratory-state specific 3D volumes using the respiratory state classes predicted by RNN. The overall pipeline is summarised in Fig. 1.

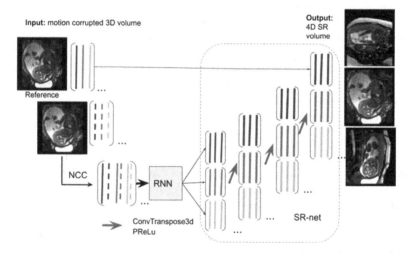

Fig. 1. Reconstruction pipeline for respiratory motion resolved 3D+t abdominal and in-utero MRI.

2.3   Self-supervised RNN

Due to sequential acquisition of slices when using the SWEEP sequence, the respiratory signal is embedded in the neighborhood slices in the arrow of acquisition time. To separate the slices into different respiratory states, we train a bi-directional self-supervised RNN (SRNN).

We first generate a reference volume based on 1D convolution with a Gaussian kernel along the acquisition axis (Z-axis). Intuitively, the reference volume is most similar to the average states of inhale and exhale. We then calculate the normalised cross-correlation (NCC) between each slice in the motion corrupted volume and the reference volume and the average inhale and exhale states are identified as the peaks of NCC sequence. We then separate those two average states based on their timing orders. The remaining states are identified linearly based on the distance between the average states. The approximate states automatically determined by this approached are then used to train bi-directional RNN.

Consider the input motion corrupted MRI scan as a group of 2D image sequence $I = \{I_1, I_2, ..., I_T\}$, where the slice number is equivalent to the arrow of time. For analysing temporal features, we use the bidirectional LSTM network to formulate the respiratory states that naturally embedded in the neighbourhood slices, where both past and future events is used for prediction [13]. Each LSTM unit computes the hidden vector sequence $h = \{h_1, h_2..., h_T\}$ and memory cell $C = \{C_1, C_2..., C_T\}$ and output vector sequence $y = \{y_1, y_2, ...y_T\}$ by bidirectional iterating from the sequence time $t = 1$ to T and $t = T$ to 1. We built a three layer bidirectional LSTM and set the total classes to 10 in the fully connected layer. In this work, we automated annotate each slice with a respiratory state, then segment the volume into multiple 20-slice subvolumes with 1 slice overlap, the input of SRNN is a 20×20 cosine similarity matrix and the output is the last slice prediction.

2.4 Super-Resolution Reconstruction (SR-net)

Deep learning based SR methods which trained on paired low resolution and SR images are reported to outperform the traditional ones [15]. Our method offer the first time non-example based SR solution, which use PSF as downgrade function and jointly penalize the MSE and TV losses. For each respiratory state, the selected slices are used to perform SR reconstruction. As shown in Fig. 1, we train a four layer 3D ConvNet with parametric rectified linear unit (PReLu), where the loss function is defined as the combined reconstruction error and the total variation (TV) regularisation [16]. As proposed previously [6], we treat PSF as a 3D Gaussian function with Full width at half maximum (FWHM) equal to the slice-thickness in the through-plane direction. The reconstruction error then can be expressed as $E(V) = \sum_{jk}(R_{jk} - S_{jk})^2$ where R_{jk} refers to the intensity of the voxel j in each selected slice indexed by k and S_{jk} are that simulated from isotropic super-resolved volume V using the PSF. The loss function of SR-net is formulated as:

$$L_{SR} = E(V) + \sum_i \lambda_i \sum_l TV^{1D}(V(d_i)_l) \tag{1}$$

where λ_i is weighting coefficient that balances the TV loss in different orientations, and $V(d_i)$ denotes every possible l-dimensional slice of V following dimension d_i. SRnet takes less than 1 min at test time for a 4D reconstruction of our data, while previously proposed methods that were build to handle randomly oriented PSFs take around 40 min [5] and 5 h [6].

2.5 Implementation Details

The method was implemented using Python and Pytorch. The network was trained in two steps, first the SRNN is trained with 1359 subvolumes from 2 subjects with 8 groups of breathing states. Then, the SRnet is trained on 24 3D volumes from 3 subjects. For both SRNN and SRnet training, we use Adam as an optimisation tool. For SRNN, the learning rate has been tested from 0.01 to 1 and set to 0.1 based on empirical results. To avoid the over-fitting, we set the weight decay to 0.01, which add L2 regularization of the weights into the optimisation procedure. For SR-net, we set the TV loss weights to 0.01, 0.01, and 0.1 to enforce the data smoothness in Z-axis. We set the total epoch to 5000. The total training time is 5 h.

3 Results

3.1 Simulated Experiment

To validate the classification accuracy of the SRNN, we generated a simulated dataset with 5 different respiratory states sampling. For a real in-utero dataset we classified slices into motion states using combination of peak selection and manual input. We then reconstructed the average motion state and registered it to the acquired slices of the other respiratory states. A breathing cycle was then simulated based on the choice of eight slices from each group with random starting state. We tested the peak selection method and the SRNN to the simulated dataset.

Table 1. Comparison respiratory state classification accuracy on the simulation dataset

Data ID	1	2	3	4	5	$Mean \pm Stdev$
Peak analysis	39.61%	41.03%	27.86%	69.30%	47.72%	$45.10\% \pm 13.69\%$
SRNN	77.81%	77.00%	77.50%	77.60%	78.62%	$77.71\% \pm 0.53\%$

Table 1 shows that SRNN achieved close to 80% accuracy for all five breathing states, while original peak selection had much lower accuracy. This was mainly due to confusion of the neighbouring classes or average inhale and exhale states.

3.2 Real Data Reconstructions

MRI data were acquired on a 3T clinical system (Achieva, Philips Healthcare, Best, Netherlands) using a 2D bSSFP sequence with the SWEEP technique [14].

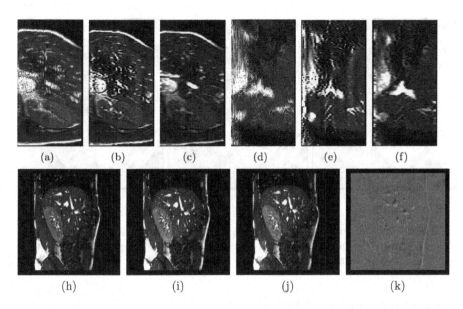

(a)   (b)   (c)   (d)   (e)   (f)

(h)   (i)   (j)   (k)

Fig. 2. Comparison reconstruction results of an abdominal subject. The top row shows the through-plane views: the original motion corrupted MRI scan (a), (d); the SVR reconstruction (b), (e); and the propose reconstruction (c), (f). The bottom row shows the Z-axis view of the original scan (h); the reconstruction results of inhale (i) and exhale state (j) and their difference (k)

Informed consent was obtained from 2 healthy adult volunteers (kidney) and 10 pregnant volunteers (gestational ages: 23–36 weeks) who were scanned in the supine position with routine blood pressure and pulse oximetry monitoring. For kidney and uterus acquisitions, the TR/TE is 5.7/2.8 and 7.3/3.6 ms, the sweep rate is 0.37 and 0.17 mm/s, and the slice thickness is 3 and 4 mm, respectively.

The reconstruction results of the abdominal scan is shown in Fig. 2. For single orientation acquisition motion artifacts are present in SVR reconstruction in spite of the automatic rejection of misaligned slices (b, e). On the other hand, SR reconstruction of slices (c, f) selected using our proposed method resolved most of the breathing artifacts. The bottom row demonstrate the proposed SRNN can accurately separate the inhale and exhale respiratory states.

Figure 3 shows a similar comparison for abdominal MRI of a pregnant patient. As highlighted in the red box, where the artifact is caused by a deep breath, due to lack of a good target with only one stack of 2D MR images, the state-of-art SVR method failed in the area with large motion corruption.

For quantitative analysis, we calculated the PSNR and SSIM between the reconstruction and sparsely selected 2D images with three different respiratory groups including average, inhale and exhale breathing states. We compared the proposed method with two state-of-the-art SVR software, the SVR [6] and NiftyMic [8]. For fair comparison, we use the selected 2D images as a target and

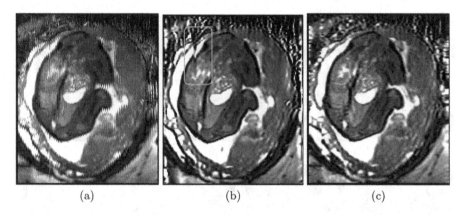

(a)   (b)   (c)

Fig. 3. Comparisons of perinatal subject with through-plane views: (a) the original motion corrupted image; (b) SVR based reconstruction; (c) the proposed reconstruction. (Color figure online)

use free-form deformation (FFD) based method in MIRTK package.[1] to register the reconstructions from SVR and NiftyMic to each respiratory group. Our 4D reconstruction results are listed as SRNN0. We then register the average state in SRNN0 to the inhale and exhale states and report the results as SRNN1 in Tables 2 and 3. The reported values in Tables 2 and 3 are the average results of 10 in-utero subjects. The results show that the proposed reconstruction pipeline can generate SR images with high fidelity to the original MRI scan.

Table 2. Comparison PSNR results between the reference volume and the proposed reconstruction pipeline with different respiratory states

	Average	Inhale	Exhale	$Mean \pm Stdev$
SVR [6]	32.82	32.48	31.59	32.30 ± 0.64
NiftyMic [8]	31.43	32.29	30.98	31.57 ± 0.67
SRNN0	36.87	36.95	35.28	36.37 ± 0.94
SRNN1	–	36.88	35.28	36.08 ± 1.13

4   Discussion and Conclusion

In this paper we proposed an efficient respiratory motion resolved 4D (3D+t) reconstruction pipeline for abdominal and in-utero MRI. We investigated the respiratory information naturally embedded in the neighborhood slices and use it to train an bidirectional RNN.

---

[1] https://mirtk.github.io/.

Table 3. Comparison SSIM results between the reference volume and the proposed reconstruction pipeline with different respiratory states

	Average	Inhale	Exhale	$Mean \pm Stdev$
SVR [6]	0.94	0.93	0.93	0.93 ± 0.01
NiftyMic [8]	0.92	0.93	0.91	0.92 ± 0.01
SRNN0	0.96	0.96	0.96	0.96 ± 0.01
SRNN1	–	0.96	0.96	0.96 ± 0.01

We propose a simple but effective motion correction and SR reconstruction pipeline for abdominal and in-utero MRI. The proposed pipeline can accurately cluster the respiratory motion of the acquired 2D images stack. The proposed self-supervised RNN utilise the NCC scores between each 2D slice and Z-axis blurred image. Such breathing motion indicator is very helpful to supervise the respiratory state clustering. The SR reconstruction stage further improves the reconstruction performances. Compared to SVRs, SRnet is a CNN pipeline that takes less than 1 min for a 4D reconstruction, while the SVR ones take around 40 min [7] and 5 h [8]. The PSNR and SSIM comparison results show that with such single orientation acquisition scenarios, the proposed pipeline with less than 20% of the sparsely selected slices outperformed the SVR methods with all the slices.

Acknowledgement. This work was supported by the National Institutes of Health Human Placenta Project [1U01HD087202-01], by the Wellcome Trust IEH Award [102431], by the Wellcome/EPSRC Centre for Medical Engineering [WT203148/Z/16/Z] and by the National Institute for Health Research (NIHR) Biomedical Research Centre at Guy's and St Thomas' NHS Foundation Trust and King's College London. The authors also thank NVIDIA Corporation for the GPU grant.

References

1. Lloyd, D.F.A., et al.: Three-dimensional visualisation of the fetal heart using prenatal MRI with motion corrected slice-volume registration. Lancet **393**, 1619–1627 (2018)
2. Story, L., Zhang, T., Aljabar, P., Hajnal, J., Shennan, A., Rutherford, M.: Magnetic resonance imaging assessment of lung volumes in fetuses at high risk of preterm birth. BJOG Int. J. Obstet. Gynaecol. **124**, 24 (2017)
3. Story, L., Hutter, J., Zhang, T., Shennan, A.H., Rutherford, M.: The use of antenatal fetal magnetic resonance imaging in the assessment of patients at high risk of preterm birth. Eur. J. Obstet. Gynecol. Reprod. Biol. **222**, 134–141 (2018)
4. Story, L., et al.: Magnetic resonance imaging assessment of lung: body volume ratios in fetuses at high risk of preterm birth. BJOG Int. J. Obstet. Gynaecol. **126**, 8 (2019)
5. Gholipour, A., Estroff, J.A., Warfield, S.K.: Robust super-resolution volume reconstruction from slice acquisitions: application to fetal brain MRI. IEEE Trans. Med. Imaging **29**(10), 1739–1758 (2010)

6. Kuklisova-Murgasova, M., Quaghebeur, G., Rutherford, M.A., Hajnal, J.V., Schnabel, J.A.: Reconstruction of fetal brain MRI with intensity matching and complete outlier removal. Med. Image Anal. **16**(8), 1550–1564 (2012)
7. Kainz, B., et al.: Fast volume reconstruction from motion corrupted stacks of 2D slices. IEEE Trans. Med. Imaging **34**(9), 1901–1913 (2015)
8. Ebner, M., et al.: An automated localization, segmentation and reconstruction framework for fetal brain MRI. In: Frangi, A.F., Schnabel, J.A., Davatzikos, C., Alberola-López, C., Fichtinger, G. (eds.) MICCAI 2018. LNCS, vol. 11070, pp. 313–320. Springer, Cham (2018). https://doi.org/10.1007/978-3-030-00928-1_36
9. Torrents-Barrena, J., et al.: Fully automatic 3D reconstruction of the placenta and its peripheral vasculature in intrauterine fetal MRI. Med. Image Anal. **54**, 263–279 (2019)
10. Ramanathan, V., Tang, K., Mori, G., Fei-Fei, L.: Learning temporal embeddings for complex video analysis. In: Proceedings of the IEEE International Conference on Computer Vision, pp. 4471–4479 (2015)
11. Fernando, B., Bilen, H., Gavves, E., Gould, S.: Self-supervised video representation learning with odd-one-out networks. In: Proceedings of the IEEE Conference on Computer Vision and Pattern Recognition, pp. 3636–3645 (2017)
12. Wei, D., Lim, J., Zisserman, A., Freeman, W.T.: Learning and using the arrow of time, pp. 8052–8060 (2018)
13. Graves, A., Schmidhuber, J.: Framewise phoneme classification with bidirectional LSTM and other neural network architectures. Neural Netw. **18**(5–6), 602–610 (2005)
14. Jackson, L.H., et al.: Respiration resolved imaging using continuous steady state multiband excitation with linear frequency sweeps. In: ISMRM, Paris, ISMRM, pp. 5–7 (2018)
15. Dong, C., Loy, C.C., He, K., Tang, X.: Image super-resolution using deep convolutional networks. IEEE Trans. Pattern Anal. Mach. Intell. **38**(2), 295–307 (2016)
16. Bredies, K., Kunisch, K., Pock, T.: Total generalized variation. SIAM J. Imaging Sci. **3**(3), 492–526 (2010)

Fast Dynamic Perfusion and Angiography Reconstruction Using an End-to-End 3D Convolutional Neural Network

Sahar Yousefi[1]($\boxtimes$), Lydiane Hirschler[1], Merlijn van der Plas[1],
Mohamed S. Elmahdy[1], Hessam Sokooti[1], Matthias Van Osch[1],
and Marius Staring[1,2]

[1] Leiden University Medical Center, Radiology, Leiden, The Netherlands
s.yousefi.radi@lumc.nl
[2] Intelligent Systems Department, Delft University of Technology,
Delft, The Netherlands

Abstract. Hadamard time-encoded pseudo-continuous arterial spin labeling (te-pCASL) is a signal-to-noise ratio (SNR)-efficient MRI technique for acquiring dynamic pCASL signals that encodes the temporal information into the labeling according to a Hadamard matrix. In the decoding step, the contribution of each sub-bolus can be isolated resulting in dynamic perfusion scans. When acquiring te-ASL both with and without flow-crushing, the ASL-signal in the arteries can be isolated resulting in 4D-angiographic information. However, obtaining multi-timepoint perfusion and angiographic data requires two acquisitions. In this study, we propose a 3D Dense-Unet convolutional neural network with a multi-level loss function for reconstructing multi-timepoint perfusion and angiographic information from an interleaved 50%-sampled crushed and 50%-sampled non-crushed data, thereby negating the additional scan time. We present a framework to generate dynamic pCASL training and validation data, based on models of the intravascular and extravascular te-pCASL signals. The proposed network achieved SSIM values of 97.3 ± 1.1 and 96.2 ± 11.1 respectively for 4D perfusion and angiographic data reconstruction for 313 test data-sets.

Keywords: Pseudo-continuous arterial spin labeling (pCASL) ·
Hadamard time-encoded ASL · Convolutional neural network (CNN) ·
4D magnetic resonance angiography (MRA) · 4D perfusion · MRI
reconstruction

1 Introduction

Arterial spin labeling (ASL) is a non-invasive MRI technique which uses magnetically labeled blood water as an endogenous tracer for assessing cerebral blood flow (CBF) [1]. Hadamard time-encoded(te)-ASL is a time-efficient approach which provides the possibility to combine the superior SNR of ASL to acquire data at different inflow times to obtain dynamic ASL-data [2]. When Hadamard

© Springer Nature Switzerland AG 2019
F. Knoll et al. (Eds.): MLMIR 2019, LNCS 11905, pp. 25–35, 2019.
https://doi.org/10.1007/978-3-030-33843-5_3

te-ASL is done with and without flow-crushing, 4D magnetic resonance angiography (MRA) and arterial input function (AIF) measurements can be obtained next to the perfusion scans [3]. While this approach improves quantification and enhances information content, it is a factor two slower, since both crushed and non-crushed data need to be acquired. Accelerating te-ASL quantification can be done either by acquiring sub-sampled data in k-space or by reducing the rank of the Hadamard matrix. However, these methods can end up reducing image quality and/or signal-to-noise (SNR) ratio.

In this work, we propose an end-to-end 3D convolutional neural network (CNN) for the reconstruction of multi-timepoint 4D MRA and perfusion scans by using half-sampled crushed as well as half-sampled non-crushed Hadamard te-ASL scans, to maintain image quality and provide accurate CBF quantification. Recently, CNNs have shown outstanding performance in medical imaging [4–6]. However, very few CNN reconstruction techniques have been proposed in the context of MRA and perfusion reconstruction. In [7] a U-net shape CNN for boosting SNR and resolution of ASL scans has been proposed. Guo et al. proposed a CNN based method for improving 3D perfusion image quality by the combined use of single- and multi-delay pseudo-continuous arterial spin labeling (PCASL) and an anatomical scan [8]. In Guo's study, ground truth perfusion maps were obtained by positron emission tomography scans. In [9] a temporal CNN approach was proposed for perfusion parameter estimation in stroke. The proposed CNN takes in the signals of interest (i.e., concentration-time curves and the AIF) to produce estimated perfusion parameter maps including cerebral blood volume (CBV), CBF, time-to-maximum, and mean transit time.

In this work, different from the previous works, we employ CNNs in order to accelerate the simultaneous acquisition of 4D MRA and perfusion measurements. One of the challenging issues is the different properties of the outputs of the proposed CNN since MRA is intrinsically sparse and has much more elongated structures than the smooth perfusion map. We tackle this issue by employing different weighting of the loss functions of these two output-types and by balancing extracted samples during training. The proposed CNN leverages the idea of dense blocks [10], arranging them in a typical U-shape [11]. Loop connectivity patterns in dense blocks improve the flow of gradients throughout the network and strengthen feature propagation and feature re-usability [4]. In this investigation, we compare the performance of several loss functions: mean square error (MSE), VGG-16 perceptual loss, structural similarity index (SSIM) and multi-level SSIM (ML-SSIM). The main contributions of our work are:

- To the best of our knowledge, we are the first to propose acceleration of the reconstruction of 4D MRA and perfusion images using interleaved sub-sampled crushed and non-crushed Hadamard te-ASL scans. To allow sub-sampling, we employed an end-to-end 3D CNN for decoding.
- We employed a framework for generating training and validation 4D MRA and perfusion scans by generalizing the Buxton kinetic model for a Hadamard te-ASL signal. Different from [12], we consider the kinetic arterial model to take into account the arterial compartment.

– We propose a CNN with a multi-level loss function and compare the proposed method with several loss functions, i.e. MSE, VGG-16 perceptual loss, and SSIM.

2 Proposed Approach

2.1 Problem

Reconstruction of dynamic perfusion scans at $H-1$ time points can be performed by the decoding of crushed te-pCASL scans of a Hadamard matrix of rank H [2]. Reconstruction of dynamic MRA scans at $H-1$ time points, next to the 4D perfusion data, is performed by the decoding of non-crushed te-pCASL scans of a Hadamard matrix of rank H and subtraction of the perfusion data from that [3]. This process can be formulated as

$$M\left(\{I_i^{\mathcal{NC}}\}, \{I_i^{\mathcal{C}}\}\right)_{i=1}^{H} = \{\mathbb{P}(t), \mathbb{A}(t)\}_{t=1}^{H-1}, \tag{1}$$

in which M is the decoding and subtraction function as described earlier [2,3], $I_i^{\mathcal{NC}}$ and $I_i^{\mathcal{C}}$ are the acquired scans of the ith row of non-crushed and crushed Hadamard te-pCASL datasets, $\mathbb{P}$ and $\mathbb{A}$ denote perfusion and angiography scans respectively.

As mentioned before, obtaining 4D-MRA and perfusion scans require two acquisitions. To accelerate this process with a factor of two, we propose an end-to-end 3D CNN, M', which reconstructs 4D-MRA and perfusion data by using interleaved half sampled crushed and half sampled non-crushed Hadamard te-pCASL scans. Therefore, the problem of reconstruction can be re-defined by

$$M'\left(\{I_{2\times i-1}^{\mathcal{NC}}, I_{2\times i}^{\mathcal{C}}\}\right)_{i=1}^{H/2} = \{\mathbb{P}(t), \mathbb{A}(t)\}_{t=1}^{H-1}. \tag{2}$$

2.2 Proposed Network

Figure 1 illustrates the proposed network, which takes 50% sub-sampled crushed and 50% sub-sampled non-crushed interleaved ASL data as input and outputs dynamic MRA and perfusion scans. For managing GPU memory, the network was implemented patch-based. The input patches (of size 53^3) are extracted from 50% sub-sampled Hadamard te-crushed and te-non-crushed scans. The outputs of the network are 14 patches of size 39^3 containing perfusion and angiography patches, each at seven different time-points. In this study, we considered a Hadamard matrix of rank 8, so the inputs are 8 patches in total (4 crushed, 4 non-crushed). In each dense block two ($3 \times 3 \times 3$) conv-BN-leaky ReLu and one ($1 \times 1 \times 1$) conv-BN-leaky ReLu, as a bottleneck layer, are stacked. Loop connectivity patterns in dense blocks are employed to improve the flow of gradients [10]. The bottleneck layers are used to increase the number of feature maps in a tractable fashion, which make the training process easier while leading to a more compact model. A down-sampling unit is followed by one $2 \times 2 \times 2$ max-pooling layer with a stride of $2 \times 2 \times 2$. In order to solve the well-known checkerboard

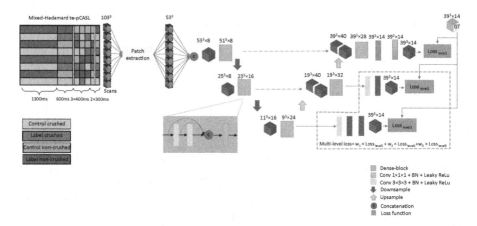

Fig. 1. Proposed network with single- and multi-level loss functions. The training data contains 50% sub-sampled and interleaved Hadamard-crushed and non-crushed scans. For the single- and multi-level loss functions, $Loss_{level1}$ and $\sum_{i=1}^{3} Loss_{leveli}$ are considered respectively. GT stands for ground truth, which is the set of angiographic and perfusion data reconstructed by the standard full-sampled decoding approach [2].

issue of the conv-transpose layer, for the up-sampling layer the feature maps are re-sized by a constant trilinear resize convolution kernel, similar to [13]. In this work we investigate the impact of several loss functions for the defined problem: MSE, which is the $L2$-norm, VGG-16 perceptual loss [14], SSIM which is composed of luminance, contrast and structural error. Later it is shown that among the mentioned loss functions, SSIM has a higher performance in terms of SSIM metric value. Therefore, we propose ML-SSIM, which is calculated based on weighting the SSIM loss function for different levels of the network, see Fig. 1.

2.3   Dataset Generation

In pCASL the arterial spins are magnetically labeled with a radiofrequency inversion pulse applied below the imaging slices in the neck vessels. The labeled blood then travels via the arteries towards the brain tissue, where they pass from the capillary compartment into the extravascular compartment. After a certain delay time after labeling which is known as the post-labeling delay (PLD) a so-called labeled image is acquired. A control image is acquired without prior labeling and by subtraction of these two images, the perfusion image can be generated. For the Hadamard te-pCASL technique, the labeling module (the typical duration of 3–4 s) is divided into several blocks (sub-boli) and a Hadamard matrix is used to determine whether a block will be played-out in label or control condition. For each voxel the Hadamard te-pCASL signal can contain both perfusion signal as well as label still residing in the arteries, i.e. angiography signals.

Since it is difficult to acquire substantial amounts of real data, we propose to model the input data, allowing to generate a sufficient amount of training data.

The ground truth output data is created by decoding fully sampled Hadamard te-ASL crushed and non-crushed data [2]. For this purpose, we create datasets based upon a tracer kinetic model for the Hadamard time-encoded pCASL signal that describes the signal a function of arterial arrival time (AAT), bolus arrival time (BAT) and CBF. In this study, for calculating the signal, the AAT and BAT information are obtained from in vivo data. The CBF maps are taken from the BrainWeb dataset by assigning CBF-values to white matter (WM), gray matter (GM) and cerebrospinal fluid (CSF).

Figure 2 shows the proposed framework for synthetically generating training and validation datasets. For this goal, we leverage the well-known Buxton kinetic model [15], which has been defined for normal ASL, and defined a tracer delivery function (for tissue voxels and arteries) and a tracer accumulation (perfusion) function for each bolus of Hadamard encoded labeling scheme. The final kinetic model is then generated by performing the convolution of the AIF and the residue function. Equations (3) and (4) define the obtained model for large arteries and tissue signals for a Hadamard scheme of 8 encoding steps respectively. These equations can be generalized for Hadamard matrices of higher rank.

$$
S_{artery} = \begin{cases} 0 & \text{if } t < \Delta t_b \\ M_{0a} \cdot aCBV \cdot L_r(b) \times e^{\frac{-\Delta t_b}{T_{1b}}} & \text{if } \Delta t_b + \sum_{b'=1}^{b-1} \tau_{b'} \le t < \Delta t_b + \sum_{b'=1}^{b} \tau_{b'} \\ 0 & \text{if } t \ge \Delta t_b + \sum_{b'=1}^{N} \tau_{b'} \end{cases}
\tag{3}
$$

$$
S_{tissue} = \begin{cases} 0 & \text{if } t < \Delta t_b \\ \gamma \Gamma_{\beta=0} & \text{if } \Delta t_a \le t < \Delta t_a + \tau_1 \\ \gamma [\Gamma_{\beta=1} + \Xi_{1:1}] & \text{if } \Delta t_a + \tau_1 \le t < \Delta t_a + \sum_{b=1}^{2} \tau_b \\ \gamma [\Gamma_{\beta=B-1} + \Xi_{B-1:1}] & \text{if } \Delta t_a + \sum_{b=1}^{B-1} \tau_b \le t < \Delta t_a + \sum_{b=1}^{B} \tau_b; B \in [3,7] \\ \gamma \Xi_{N:1} & \text{if } t \ge \Delta t_a + \sum_{b=1}^{N} \tau_b; N = 7 \end{cases}
\tag{4}
$$

in which τ_b is label duration for the bth sub-bolus, N is the number of sub-boluses, M_{0a} is the magnetization of arterial blood, Δt_a is AAT which represents the arrival time of the labeled blood in the artery, Δt_b is BAT which represents the arrival time of labeled blood in the tissue, T_{1a} is the arterial blood relaxation time, f is CBF (millimeter per gram per second), κ is static tissue signal, aCBV is arterial cerebral blood volume,

$$
\gamma = M_{0a} \cdot f \cdot e^{\frac{-\Delta t_a}{T_{1a}}} \cdot T_{1a},
\tag{5}
$$

$$
\Gamma_\beta = L_r(\beta+1) \left(1 - e^{-\frac{t-\Delta t_a - \sum_{b=1}^{\beta} \tau_b}{T_{1a}}} \right),
\tag{6}
$$

and

$$
\Xi_{\beta:\beta'} = \sum_{b'=\beta}^{\beta'} L_r(b') \left(e^{-\frac{t-\Delta t_a - \sum_{b=1}^{b'} \tau_b}{T_{1a}}} - e^{-\frac{t-\Delta t_a - \sum_{b=1}^{b'-1} \tau_b}{T_{1a}}} \right).
\tag{7}
$$

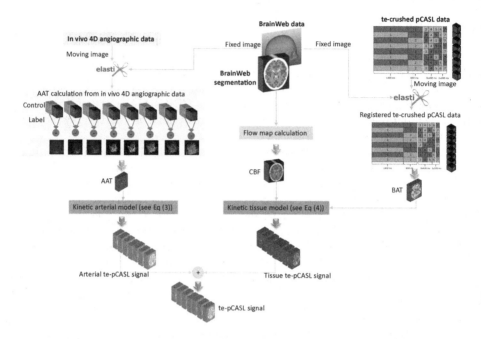

Fig. 2. Data generator framework for one subject, the inputs of the framework (shown in bold) are: in vivo information (include BAT, AAT, and CBF) and anatomical data from BrainWeb [16], the outputs are pCASL scans which by decoding the perfusion and angiographic maps are obtained (see Sect. 2.3), Left: a map of the arrival time of the label in arteries ('arterial arrival time' or AAT) is created from high resolution 4D MRA ASL scan (Cinema, 8 time-points with 200 ms temporal resolution, and a spatial resolution of $0.82 \times 0.82 \times 1.02$ mm) and the data are registered to a subject from BrainWeb dataset, and leads to an AAT map for the subject, then the AAT map is fed into Eq. (3) to calculate the kinetic arterial model. Middle pipeline: using the GM and WM segmentation of the subject and assigning literature values to the flow map (or CBF) of GM and WM the flow map is calculated, which serves as one of the inputs to the kinetic model of the tissue (Eq. 4). Right: an in vivo te-crushed pCASL data is registered to the same subject of BrainWeb and by Hadamard decoding the registered data, the arrival of the label at tissue level ('bolus arrival time', or BAT-map) is calculated. This serves as the other input to the kinetic model of the tissue (Eq. 4). The tissue signal and arterial signal are summed together to form the te-pCASL.

$L_r(b)$ is 0 if the bth bolus in the rth row is control, and it is 1 if the bth bolus in the rth row is label. For voxels containing large arteries, the pCASL signal can be computed by $S_{voxel} = S_{tissue} + S_{artery}$. The calculated tracer kinetic model is a function of AAT, BAT and CBF. In this study, the anatomic structures are obtained from the BrainWeb database [16]. In order to obtain the tracer signal, the AAT and BAT and blood maps are extracted from in vivo data, then registered with a subject from the BrainWeb dataset by Elastix [17]. The ground truth, i.e. 4D MRA and perfusion scans, are obtained by normal Hadamard decoding of the pCASL images [18]. To evaluate the generated data, the signal evolution pattern was validated by the Buxton curve model [15].

In this study, the dataset was generated for a Hadamard-8 matrix with seven blocks of respectively 1300, 600, 400, 400, 400, 300 and 300 ms with an additional 265 ms delay before the start of readout. Using the permuted in vivo information (include 6 BAT, 4 AAT) and registering those with the anatomical information from the BrainWeb dataset (consisting of 20 normal subjects and CBF) and calculating the Hadamard te-pCASL (Eqs. (3) and (4)), this study contains 1564 distinct simulated data-sets each including crushed and non-crushed input data for 8 Hadamard-encodings. By decoding each of the generated crushed and non-crushed te-pCASL data, the corresponding angiographic and perfusion output data at 7-time points, as the ground truth, are obtained. The scans were divided into 1096 subjects for training, 155 for validation and 313 for testing.

3 Experimental Results

We implemented the proposed networks in Google's Tensorflow. The patch extraction was done parallel and randomly using a multi-threaded daemon process on the CPU and then patches were fed to the network on the GPU during the training process. To tackle the sparsity of MRA with respect to the perfusion scans, 75% of the patches were extracted from the region containing arteries.

Table 1. Comparison of the different networks for perfusion and angiographic images (in gray the perfusion and in white the angiographic results), the best results for perfusion and angiography are shown in blue and green respectively, PL stands for perceptual loss. A Wilcoxon signed-rank test is performed between ML-SSIM and other loss functions for perfusion and angiography, where † indicates a statistically significant difference with $p < 0.05$.

Loss function		SSIM%	MSE	SNR	pSNR	# of param
MSE	$(\mu \pm \sigma)$	$97.0 \pm 1.1^{\dagger}$	$0.02 \pm 0.02^{\dagger}$	$3.84 \pm 1.41^{\dagger}$	$30.8 \pm 2.3^{\dagger}$	169,152
	med	96.8	0.02	4.37	30.7	
	$(\mu \pm \sigma)$	$70.0 \pm 45.5^{\dagger}$	$0.45 \pm 2.31^{\dagger}$	$1.28 \pm 0.24^{\dagger}$	$33.8 \pm 22.0^{\dagger}$	
	med	99.5	0.17	1.31	46.9	
SSIM	$(\mu \pm \sigma)$	$96.3 \pm 1.7^{\dagger}$	$0.04 \pm 0.04^{\dagger}$	$2.97 \pm 1.26^{\dagger}$	$28.9 \pm 1.6^{\dagger}$	169,152
	med	96.99	0.03	3.49	28.44	
	$(\mu \pm \sigma)$	96.7 ± 12.5	$0.37 \pm 2.64^{\dagger}$	$1.15 \pm 0.70^{\dagger}$	$33.8 \pm 22.1^{\dagger}$	
	med	98.9	0.20	1.12	46.28	
PL	$(\mu \pm \sigma)$	$96.1 \pm 1.8^{\dagger}$	$0.05 \pm 0.39^{\dagger}$	$5.14 \pm 1.81^{\dagger}$	$33.7 \pm 3.6^{\dagger}$	169,152
	med	96.29	0.01	6.00	33.19	
	$(\mu \pm \sigma)$	$96.3 \pm 12.8^{\dagger}$	$0.16 \pm 2.23^{\dagger}$	$3.03 \pm 3.21^{\dagger}$	$38.4 \pm 25.3^{\dagger}$	
	med	99.8	0.04	2.30	51.34	
ML-SSIM	$(\mu \pm \sigma)$	97.3 ± 1.1	0.03 ± 0.15	6.18 ± 2.38	35.0 ± 3.2	181,692
	med	97.1	0.01	7.48	34.63	
	$(\mu \pm \sigma)$	96.2 ± 11.1	0.44 ± 3.17	1.67 ± 0.52	35.4 ± 23.2	
	med	99.7	0.10	1.76	49.35	

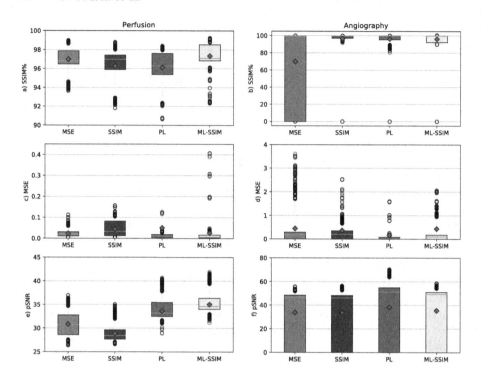

Fig. 3. Boxplots for different metrics and CNNs for perfusion and angiographic results, PL stands for VGG-16 perceptual loss. For (e) a few outliers smaller than 25 and for (c)/(d) a few outliers larger than 0.45/4 are not shown.

The input patches were augmented by white noise extracted from a Gaussian distribution with zero mean and random standard deviation between 0 and 5, left-to-right flipping, and random rotation (up to $\pm 18°$).

Evaluation of the proposed networks has been performed by calculating SSIM, MSE, SNR, and peak signal to noise ratio (pSNR), comparing the ground truth reconstruction using full sampling with that of the neural network using 50% subsampling. Table 1 tabulates a quantitative comparison between the mentioned loss functions. A statistically significant difference (with $p < 0.05$) between ML-SSIM and all the other methods, for perfusion and angiography, can be observed.

Figure 3 depicts the boxplots of the metrics on the test set. The network using the ML-SSIM loss function had a value of 97.3 ± 1.1, 6.2 ± 2.4 and 35.0 ± 3.2 for SSIM, SNR and pSNR respectively, and the best performance for perfusion reconstruction. The network with the SSIM loss function had a SSIM of 96.7 ± 12.5 for angiography reconstruction, i.e. the best performance in terms of SSIM while it does not show a statistically significant difference from the network with the ML-SSIM loss function. Also for angiography reconstruction the network with perceptual loss had the best value for SNR and pSNR while the network with the ML-SSIM loss function with the values of 1.67 ± 0.52 and 35.4 ± 23.2 for SNR and pSNR respectively had the second rank.

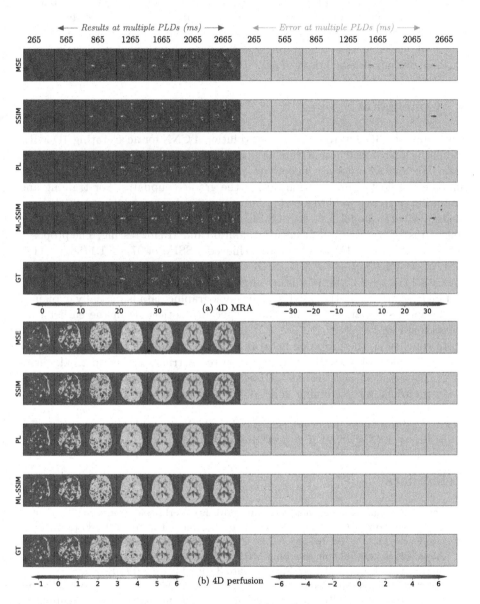

Fig. 4. Qualification results, 4D (a) MRA and (b) perfusion and error at multiple time points after labeling of arterial spins in a single-slice for the different networks, GT stands for ground truth which is obtained from fully sampled decoding and subtracting (see Eq. 1).

Figure 4 exemplifies the qualification compassion for 4D MRA and perfusion reconstruction between the different CNNs. The lower SNR and higher pSNR variance for the angiographic data is partially explained by the intrinsic sparsity of that data, especially noticeable in the earlier time points, see Fig. 4a.

It takes an average of 205 ± 232 ms from the ML-SSIM network to reconstruct all perfusion and angiography scans from the interleaved sparsely-sampled crushed and non-crushed data of size 107^3.

4   Conclusion

We proposed a 3D end-to-end fully convolutional CNN for accelerating 4D MRA and perfusion reconstruction from half-sampled crushed and non-crushed pCASL data. We leveraged loop connectivity patterns in the network architecture to improve the flow of information during the gradient updates. For training and validation purposes we developed a data generator framework based on the generalized kinetic model for the pCASL signal. The generated dataset included 1096 scans for training, 155 scans for validation and 313 for testing. The proposed network with ML-SSIM loss function achieved a SSIM of $97.3 \pm 1.1/96.2 \pm 11.1$, MSE of $0.03 \pm 0.15/0.44 \pm 3.17$, SNR of $6.18 \pm 2.38/1.67 \pm 0.52$, and pSNR of $35.0 \pm 3.2/35.4 \pm 23.2$ for perfusion/angiography reconstruction. The lower SNR and higher variance in the pSNR for the angiographic data is partially explained by the intrinsic sparsity of that data, especially noticeable in the earlier time points, see Fig. 4a.

In conclusion, the proposed network obtained promising results for the challenging problem of 4D MRA and perfusion reconstruction. The method, therefore, may assist an accelerated MRI scanning workflow. A further step of this study is enriching the training and validation datasets with in vivo data.

Acknowledgements. This work is financed by the Netherlands Organization for Scientific Research (NWO), VICI project 016.160.351.

References

1. Ferlay, J., et al.: Cancer incidence and mortality worldwide: sources methods and major patterns in GLOBOCAN 2012. Int. J. Cancer **136**(5), 359–386 (2015)
2. van Osch, M.J.P., Teeuwisse, W.M., Chen, Z., Suzuki, Y., Helle, M., Schmid, S.: Advances in arterial spin labelling MRI methods for measuring perfusion and collateral flow. J. Cereb. Blood Flow Metab. **38**(9), 1461–1480 (2018)
3. Petersen, E.T., Mouridsen, K., Golay, X.: The QUASAR reproducibility study, part II: results from a multi-center arterial spin labeling test-retest study. Neuroimage **49**(1), 104–113 (2010)
4. Yousefi, S., et al.: Esophageal gross tumor volume segmentation using a 3D convolutional neural network. In: Frangi, A.F., Schnabel, J.A., Davatzikos, C., Alberola-López, C., Fichtinger, G. (eds.) MICCAI 2018. LNCS, vol. 11073, pp. 343–351. Springer, Cham (2018). https://doi.org/10.1007/978-3-030-00937-3_40

5. Elmahdy, M.S., et al.: Robust contour propagation using deep learning and image registration for online adaptive proton therapy of prostate cancer. Med. Phys. **46**, 3329–3343 (2019)
6. Gong, K., et al.: Iterative pet image reconstruction using convolutional neural network representation. TMI **38**(3), 675–685 (2018)
7. Gong, E., Pauly, J., Zaharchuk, G.: Boosting SNR and/or resolution of arterial spin label (ASL) imaging using multi-contrast approaches with multi-lateral guided filter and deep networks. In: Proceedings of the Annual Meeting of the International Society for Magnetic Resonance in Medicine, Honolulu, Hawaii (2017)
8. Guo, J., Gong, E., Goubran, M., Fan, A., Khalighi, M., Zaharchuk, G.: Improving perfusion image quality and quantification accuracy using multi-contrast MRI and deep convolutional neural networks. In: ISMRM, Paris, France (2018)
9. Ho, K.C., Scalzo, F., Sarma, K.V., El-Saden, S., Arnold, C.W.: A temporal deep learning approach for MR perfusion parameter estimation in stroke. In: 23rd ICPR, pp. 1315–1320. IEEE (2016)
10. Huang, G., Liu, Z., Van Der Maaten, L., Weinberger, K.Q.: Densely connected convolutional networks. In: CVPR, pp. 4700–4708 (2017)
11. Jégou, S., Drozdzal, M., Vazquez, D., Romero, A., Bengio, Y.: The one hundred layers tiramisu: fully convolutional densenets for semantic segmentation. In: CVPR, pp. 11–19 (2017)
12. Zhao, L., Fielden, S.W., Feng, X., Wintermark, M., Mugler III, J.P., Meyer, C.H.: Rapid 3D dynamic arterial spin labeling with a sparse model-based image reconstruction. Neuroimage **121**, 205–216 (2015)
13. Dong, C., Loy, C.C., He, K., Tang, X.: Image super-resolution using deep convolutional networks. TPAMI **38**(2), 295–307 (2015)
14. Johnson, J., Alahi, A., Fei-Fei, L.: Perceptual losses for real-time style transfer and super-resolution. In: Leibe, B., Matas, J., Sebe, N., Welling, M. (eds.) ECCV 2016. LNCS, vol. 9906, pp. 694–711. Springer, Cham (2016). https://doi.org/10.1007/978-3-319-46475-6_43
15. Buxton, R.B., Frank, L.R., Wong, E.C., Siewert, B., Warach, S., Edelman, R.R.: A general kinetic model for quantitative perfusion imaging with arterial spin labeling. MRM **40**(3), 383–396 (1998)
16. Cocosco, C.A., Kollokian, V., Kwan, R.K.-S., Pike, G.B., Evans, A.C.: Brainweb: online interface to a 3D MRI simulated brain database. NeuroImage **5**, 425 (1997)
17. Klein, S., Staring, M., Murphy, K., Viergever, M.A., Pluim, J.P.: Elastix: a toolbox for intensity-based medical image registration. TMI **29**(1), 196–205 (2010)
18. Hirschler, L., et al.: Transit time mapping in the mouse brain using time-encoded pCASL. NMR Biomed. **31**(2), e3855 (2018)

APIR-Net: Autocalibrated Parallel Imaging Reconstruction Using a Neural Network

Chaoping Zhang[1]([⊠]), Florian Dubost[1], Marleen de Bruijne[1,2], Stefan Klein[1], and Dirk H. J. Poot[1]

[1] Biomedical Imaging Group Rotterdam, Erasmus MC, Rotterdam, The Netherlands
c.zhang@erasmusmc.nl
[2] Department of Computer Science, University of Copenhagen, Copenhagen, Denmark

Abstract. Deep learning has been successfully demonstrated in MRI reconstruction of accelerated acquisitions. However, its dependence on representative training data limits the application across different contrasts, anatomies, or image sizes. To address this limitation, we propose an unsupervised, auto-calibrated k-space completion method, based on a uniquely designed neural network that reconstructs the full k-space from an undersampled k-space, exploiting the redundancy among the multiple channels in the receive coil in a parallel imaging acquisition. To achieve this, contrary to common convolutional network approaches, the proposed network has a decreasing number of feature maps of constant size. In contrast to conventional parallel imaging methods such as GRAPPA that estimate the prediction kernel from the fully sampled autocalibration signals in a linear way, our method is able to learn nonlinear relations between sampled and unsampled positions in k-space. The proposed method was compared to the start-of-the-art ESPIRiT and RAKI methods in terms of noise amplification and visual image quality in both phantom and in-vivo experiments. The experiments indicate that APIR-Net provides a promising alternative to the conventional parallel imaging methods, and results in improved image quality especially for low SNR acquisitions.

Keywords: Magnetic resonance imaging · Reconstruction · Parallel imaging · Neural network

1 Introduction

Magnetic resonance imaging (MRI) provides versatile contrast information for clinical diagnosis. However, its long scan time remains a limitation. To reduce scan time, parallel imaging [2,5] has been proposed to reconstruct subsampled k-spaces acquired by multi-channel coils and is widely used in clinic. Recently, deep learning was also demonstrated to enable fast imaging with the reconstruction model trained on representative data [3,9,10].

© Springer Nature Switzerland AG 2019
F. Knoll et al. (Eds.): MLMIR 2019, LNCS 11905, pp. 36–46, 2019.
https://doi.org/10.1007/978-3-030-33843-5_4

Despite the current success of deep learning in MRI reconstruction, most methods are size, contrast, or anatomy specific. Also they depend on the corresponding training data, and may create inaccurate reconstruction for features not seen in training data. Recurrent inference machines have been introduced to iteratively reconstruct heterogeneous raw MRI data with different anatomies and acquisition settings [4]. However, training data is still needed and influences reconstruction performance. A database-free deep learning approach for fast imaging (RAKI) was proposed for parallel imaging reconstruction [1]. It learns the prediction kernel with an artificial neural network from fully sampled autocalibration signals (ACS) and subsequently uses the learned kernel to predict the unsampled signals. In this method, the nonlinear estimation of the prediction kernel enables improved noise resilience compared to the linear GRAPPA method.

In this work, we propose a different unsupervised k-space completion method for parallel imaging, called Autocalibrated Parallel Imaging Reconstruction using a Neural Network (APIR-Net). Contrary to RAKI which as a 2D method predicts the unsampled signals for a 2D k-space using prediction kernels learned from the ACS signals, APIR-Net predicts all signals of a 3D full k-space from the subsampled k-space utilizing all sampled signals, including ACS signals and beyond.

Most image based neural network architectures use downsampling (and subsequent upsampling) operators with increasing number of feature maps to force the network to use higher level image features. This assumes that such high level features are present at rather small scales in the images. In k-space small scale features represent large scale image features and hence such high level features are less likely to be present, yet preservation of small scale information is essential. On the other hand, for MRI using a multi-channel receive coil, signal redundancy exists among the channels. Hence, inspired by, but in contrast to, the U-net architecture, APIR-Net decreases the number of feature maps while preserving their size throughout all layers.

To improve the computational efficiency, we propose to train APIR-Net in a hierarchical process, starting from a small portion of k-space in the center region until the full size k-space. The network trained at a lower level provides initialized weights for a subsequent higher level's training. The performance of

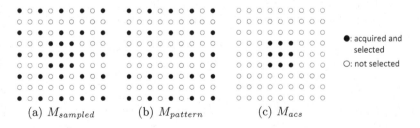

(a) $M_{sampled}$ (b) $M_{pattern}$ (c) M_{acs}

Fig. 1. An example of the masks of all sampled positions $M_{sampled}$, regularly subsampled positions $M_{pattern}$, and the ACS positions M_{acs} in PE directions.

APIR-Net was evaluated with phantom and in-vivo acquisitions with comparison to GRAPPA [2], ESPIRiT [7], and RAKI [1] methods.

2   Methods

2.1   Conventional Parallel Imaging Reconstruction [2]

In a parallel imaging acquisition with a multi-channel receive coil, k-space is subsampled regularly with a small fully sampled ACS region in k-space center. We represent the masks of all sampled positions, regularly subsampled positions, and the ACS positions in phase encoding (PE) directions as $M_{sampled}$, $M_{pattern}$, and M_{acs}, respectively, as illustrated in Fig. 1. In GRAPPA, the unsampled signals are predicted as

$$S_{predicted} = n \circledast (S \circ M_{pattern}), \tag{1}$$

where $\circledast$ represents convolution operation and $\circ$ represents pixelwise multiplication. S represents all signals in k-space. n is the prediction kernel that is trained on the ACS region M_{acs} by the least squares fitting as

$$\hat{n} = \arg\min_{n} \|(S - n \circledast (S \circ M_{pattern})) \circ M_{acs}\|_2^2 + \lambda \|In\|_2^2, \tag{2}$$

where λ is a scalar and I is the identity matrix. The second term in Eq. 2 represents Tikhonov regularization, and actual implementations may add terms with shifted versions of $M_{pattern}$.

The final interpolated k-space is computed as

$$S_{final} = S \circ M_{sampled} + S_{predicted} \circ (1 - M_{sampled}), \tag{3}$$

and the final image X is computed as

$$X = \sqrt{\frac{1}{C} \sum_c |iFFT_c(S_{final})|^2}, \tag{4}$$

where C is the number of channels and $iFFT_c$ is the 3D inverse Fourier transform on channel c.

2.2   APIR-Net Reconstruction

Problem Formulation. Instead of explicitly computing the convolution kernel n and applying it over the full k-space as in GRAPPA [2] or RAKI [1], our approach recovers the full k-space by training of a deep convolutional neural network:

$$\hat{\theta} = \arg\min_{\theta} \|(A_\theta(S \circ M_{pattern}) - S) \circ M_{sampled}\|_2^2, \tag{5}$$

where A_θ is the function to predict the full k-space from $S \circ M_{pattern}$ and is parametrized by θ as a neural network. With the input equal to $S \circ M_{pattern}$, only the regularly subsampled signals are effectively fed into the network, whereas the loss is computed on all sampled positions $M_{sampled}$. The final image X is reconstructed by Eq. 4 with $S_{final} = A_{\hat{\theta}}(S \circ M_{pattern})$.

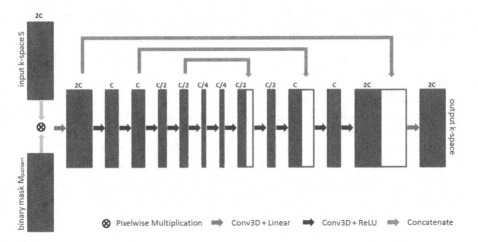

Fig. 2. The neural network architecture of APIR-Net.eps

Network Architecture. The detailed network architecture is shown in Fig. 2. The real and imaginary components of the complex k-space are concatenated in channel dimension, resulting in an input with $2C$ channels. Similarly, the output with $2C$ channels is finally converted to a C-channel complex-valued k-space. As the multi-channel k-space in a parallel imaging acquisition is highly correlated across the channel dimension, in APIR-Net, the layers have progressively reduced number of channels, or feature maps, as the depth of the encoder increases, while the size of each feature map remains unchanged. The first and last convolutional layers use a kernel size 5×5 followed by the linear activation function. The remaining convolutional layers are with kernel size of 3×3 followed by the ReLU activation function. Periodic padding is used for all convolutional layers, such that the border of the input for each convolutional layer is padded from the opposite border to maintain the size of output equal to the input.

Training. The network is trained in a hierarchical way. This is motivated by the main aspect that allows parallel imaging reconstruction: the differences in coil sensitivity of the multi-channel coil. In image domain the coil sensitivity is multiplicative in k-space this appears as translation invariant convolution. Hence, to accelerate training we start with the ACS part and progressively increase the size until the full k-space is included. Details of the training are provided in Algorithm 1.

3 Experiments

3.1 Evaluation with Phantom Acquisition

The 3D k-space of the ACR-NEMA MRI Phantom was fully acquired with fast spin echo sequence using a 3T GE Discovery MR750 scanner and an

Algorithm 1: The proposed hierarchical training method.

Input : S, $M_{sampled}$, $M_{pattern}$, $L \leftarrow$ number of levels in the hierarchical training, $k_l \in \mathbb{Z}^3 \leftarrow$ position vector indicating the k-space region being used in training of level l

Output: $\hat{\theta}$

1 Normalize magnitude $|S \circ M_{sampled}|$ to $[0,1]$
2 Initialize $\hat{\theta}_0$ randomly with a uniform distribution within $[-0.05, 0.05]$
3 **for** $l \leftarrow 1$ **to** L **do**
4      $S_{in,l} = S(k_l) \circ M_{sampled}(k_l)$, $M_{in,l} = M_{pattern}(k_l)$, $M_{out,l} = M_{sampled}(k_l)$
5      $\hat{\theta}_l = \arg\min_\theta \|(A_\theta(S_{in,l} \circ M_{in,l}) - S_{in,l}) \circ M_{out,l}\|_2^2$ starting from $\theta = \hat{\theta}_{l-1}$ until convergence
6 **end**
7 **return** $\hat{\theta} = \hat{\theta}_L$

eight-channel birdcage-like receive brain coil (8HRBRAIN). The scan parameters include repetition time $(TR) = 2800$ ms, echo train length $(ETL) = 60$, bandwidth $(BW) = 83.33$ kHz, field of view $(FOV) = 20.5 \times 20.5 \times 20.5$ cm^3, Matrix size $= 192 \times 192 \times 192$. The cylindrical phantom was placed axially in the coil array, with S/I as frequency encoding (FE) direction.

Reconstructions with GRAPPA [2], ESPIRiT [7], l_1-ESPIRiT (ESPIRiT integrating the regularization of the l_1-norm with a sparsity transform) [7] and APIR-Net were performed on a retrospectively subsampled k-space from the full acquisition with a subsampling factor of 2×2 in two PE directions. The k-space center with the size $[25 \times 25]$ in [PE1, PE2] was fully sampled as the ACS region. To highlight the noise amplification suppression capability of both methods, simulated Gaussian noise was added to the acquired positions in the subsampled k-space.

For GRAPPA, a convolution kernel size of $[5, 9, 9]$ in [FE, PE1, PE2] directions was selected from a range of options to obtain an optimal reconstruction. To fairly compare GRAPPA we reconstructed both without Tikhonov regularization $(\lambda = 0)$ as well as with a value of λ for which aliasing artifacts started to appear.

ESPIRiT reconstruction was performed using implementation from the BART toolbox [8]. In the reconstruction, the eigenvector maps for the first two eigenvalues were used. For l_1-ESPIRiT reconstruction, a regularization term of l_1-norm with wavelet transform was used. The strength of the regularization was selected towards a low noise level while avoiding visually obvious blurriness or artifacts $(r = 0.01)$.

In APIR-Net reconstruction, the training was converged with sufficient number of epochs. The settings of the hierarchical training are heuristically determined and are shown in Table 1. Adam was used as optimizer $(\beta_1 = 0.9, \beta_2 = 0.99, \epsilon = 10^{-20})$ and no regularization was used in APIR-Net.

The computation time was around 5 min for GRAPPA and around 10 min for ESPIRiT with the CPU Intel Xeon E5503, and was around 120 min for APIR-Net with the GPU NVIDIA GeForce GTX 1080Ti and the CPU Intel Core i7-8700.

Table 1. The settings for each level of the hierarchical training in APIR-Net.

	Size (phantom)	Size (in-vivo)	Initial learning rate	Number of epochs
Level 1	$32 \times 32 \times 32$	$16 \times 32 \times 32$	0.001	10000
Level 2	$48 \times 48 \times 48$	$112 \times 84 \times 64$	0.0001	5000
Level 3	$96 \times 96 \times 96$	$224 \times 164 \times 126$	0.00005	1000
Level 4	$192 \times 192 \times 192$	$224 \times 224 \times 178$	0.00005	500

The mean square error (MSE) of the phantom region in the reconstructed image with regard to the reference image was computed for each method. The reference image for MSE computation was reconstructed by the root mean squares of the inverse Fourier transform on the fully acquired k-space.

The noise amplification factor was computed with the pseudo multiple replica method [6] with 50 iterations by adding Gaussian white noise to the acquired positions in the subsampled k-space. The magnitude level of the simulated noise was the same for all replica.

3.2   Comparison to RAKI

As the current version of RAKI is a 2D method, we performed a separate experiment for comparison. The implementation of RAKI was kindly provided by the authors of [1]. To fit a 2D reconstruction method, the same 3D k-space was first fourier transformed in the FE direction to obtain k-spaces of 2D axial (PE1 × PE2) slices, which contains the most variation of the coil sensitivity. A single slice was extracted and was further subsampled by a factor of 3 in the first PE direction, with an ACS region of 25 lines. This data was reconstructed by RAKI, APIR-Net, ESPIRiT, l_1-ESPIRiT ($r = 0.01$), and GRAPPA, where (obviously) 2D Convolution kernels of otherwise identical size were used in APIR-Net (2D APIR-Net). To investigate the influence of increasing training size we additionally reconstruct with APIR-Net the 3D k-space identically subsampled by a factor 3 in the first PE direction.

3.3   Evaluation with In-Vivo Acquisitions

To evaluate the proposed method with in-vivo acquisitions, a brain scan from one volunteer with FLAIR contrast was performed with the same scanner and coil as the phantom acquisition. This study was approved by our Institutional Review Board and informed consent was obtained from the volunteer. The prospectively subsampled k-spaces skipped the corners in the PE plane. The subsampling

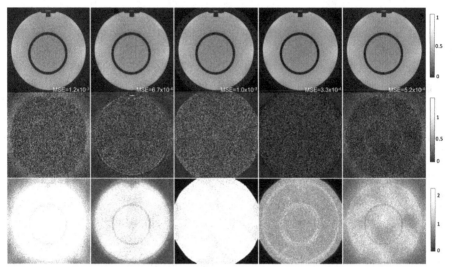

(a) GRAPPA. (b) Regularized (c) ESPIRiT. (d) l_1-ESPIRiT. (e) APIR-Net.
GRAPPA.

Fig. 3. One axial slice of the reconstructed images (first row), the reconstruction errors (second row), and the noise amplification factors (third row) of the phantom experiment.

factors are [2, 3] in [PE1, PE2] directions. The ACS region with size [25 × 25] in [PE1, PE2] was additional fully acquired as well. Other settings include $TR = 5000\,ms$, inversion time $(TI) = 1700\,ms$, $ETL = 60$, $FOV = 22.4 \times 22.4 \times 17.8\,cm^3$, Matrix size $= 224 \times 224 \times 178$, $BW = 41.67\,kHz$. The effective scan time was 3.95 min.

Similar to the experiments with phantom data, Gaussian noise was added to the acquired positions in k-spaces to investigate the noise amplification suppression capability of the methods. The settings of the hierarchical training for APIR-Net are shown in Table 1 as well. For APIR-Net reconstruction, prediction using network weights trained on different levels of the hierarchical training was also performed. Besides APIR-Net, reconstructions with GRAPPA, ESPIRiT (eigenvectors of the first two eigenvalues used), and l_1-ESPIRiT ($r = 0.01$) were also performed.

4   Results

4.1   Evaluation with Phantom Acquisition

The images of the phantom reconstructed by all methods are shown in the first row in Fig. 3. The reconstruction errors, i.e., the absolute difference between the reconstructed images and the reference image, are shown in the second row in Fig. 3. When using regularization with GRAPPA, SNR increases but

(a) GRAPPA. (b) Regularized (c) ESPIRiT. (d) l_1-ESPIRiT.
GRAPPA.

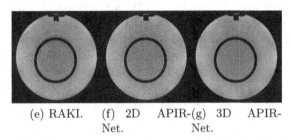

(e) RAKI. (f) 2D APIR-(g) 3D APIR-
Net.   Net.

Fig. 4. (a-f) 2D k-space reconstructions of one axial slice, (g) was reconstructed from a 3D k-space.

aliasing artifacts start to appear. APIR-Net reconstruction shows higher SNR than ESPIRiT reconstruction and the GRAPPA reconstruction without regularization, and less aliasing artifacts than the regularized GRAPPA reconstruction while having higher SNR. By integrating a properly weighted regularization term of l_1-norm of wavelet coefficients of the reconstructed image, l_1-ESPIRiT reduced noise level of ESPIRiT without raising visually obvious artifacts, and achieves the optimal image quality overall. MSEs of the reconstructed images are shown in the reconstructed images in Fig. 3. APIR-Net reconstruction shows a lower MSE than the other methods except l_1-ESPIRiT. While l_1-ESPIRiT outperforms APIR-Net in the reconstruction quality, the same regularization, i.e. the sparsity constraint in wavelet transform, of l_1-ESPIRiT can be integrated in APIR-Net as well to improve its reconstruction quality.

As shown in the third row in Fig. 3, with regularization, the noise amplification was reduced in GRAPPA reconstruction, but still clearly higher than APIR-Net reconstruction. l_1-ESPIRiT overall shows the optimal noise amplification suppression with a substantial improvement over ESPIRiT.

4.2 Comparison to RAKI

As shown in Fig. 4, RAKI increased SNR substantially compared to GRAPPA. Compared to regularized GRAPPA, with similar amount of aliasing artifacts, RAKI still achieved higher SNR. With slightly lower SNR than RAKI and higher than regularized GRAPPA, 2D APIR-Net achieved less aliasing artifacts and visually better image quality. l_1-ESPIRiT substantially increased SNR compared to ESPIRiT, and achieved visually better quality than the previous ones. With more signals available for training, 3D APIR-Net shows improvement in both SNR and artifacts than 2D APIR-Net and also better image quality than other 2D methods.

4.3   Evaluation with In-Vivo Acquisitions

The reconstructed in-vivo images are shown in Fig. 5. Regularized GRAPPA reconstruction shows reduced noise compared to GRAPPA without regularization, but aliasing artifacts appear. APIR-Net reconstruction achieves better performance than GRAPPA in both noise and aliasing artifacts. Compared to ESPIRiT, APIR-Net reduced the noise level of the image, though l_1-ESPIRiT achieves a further reduced noise level with slight blurring appears.

Figure 6 shows the images reconstructed using weights trained from different levels in the hierarchical training. The image quality (in terms of noise and aliasing artifacts) is improved with the fine tuning of higher levels training.

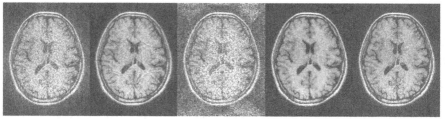

(a) GRAPPA. (b)   Regularized (c) ESPIRiT. (d) l_1-ESPIRiT. (e) APIR-Net.
GRAPPA.

Fig. 5. One axial slice of the reconstructed images of the prospectively subsampled in-vivo acquisition.

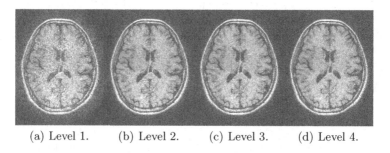

(a) Level 1.   (b) Level 2.   (c) Level 3.   (d) Level 4.

Fig. 6. One axial slice of the reconstructed in-vivo images by APIR-Net. Reconstruction using weights of the first level training (a), second level (b), third level (c), and fourth level (d).

5   Discussion and Conclusion

This work presented a novel method, APIR-Net, to reconstruct images from parallel imaging acquisitions using a neural network. While maintaining the nonlinear optimization capability of deep learning based reconstruction methods, APIR-Net does not need representative training data of additional subject scans. This enables flexibility of APIR-Net in image size, contrast, or anatomy.

Compared to GRAPPA, which estimates the unsampled signals in a linear way, APIR-Net achieves better noise amplification suppression, and thus better image quality and SNR. In GRAPPA noise amplification can be reduced by including Tikhonov regularization. However, this may introduce artifacts. In our results the regularized GRAPPA reconstruction had both stronger artifacts and higher noise level than APIR-Net. Compared to ESPIRiT, where excessive noise amplification exists, APIR-Net shows improvement of the image quality with better SNR. l_1-ESPIRiT, which integrates the assumption of sparsity in the wavelet domain, substantially improves SNR and achieves (slightly) better image quality than APIR-Net, though tuning of the regularization strength is needed. As currently APIR-Net does not use any image based prior information, we hypothesise that the APIR-Net results can be further improved by including such prior information into the reconstruction, e.g. by adding the l_1 norm of the wavelet transform of the reconstructed image as additional cost term in Eq. 5. The RAKI method [1] which trains a convolutional neural network from ACS to predict the unsampled signals showed better results than GRAPPA. In APIR-Net, which uses a substantially different network architecture and extends RAKI in that all sampled signals (including signals beyond the ACS region) are used in prediction, improved image quality is achieved.

Although it achieves the improved image quality, the current computation time of APIR-Net is much longer than GRAPPA. The high levels of the hierarchical training for APIR-Net are with typically very large size inputs (multichannel high resolution 3D k-space), which makes it computationally expensive. We expect that by using patch generation techniques and stochastic optimization, the computation time can be reduced substantially. Additionally, initialization of the network weights might be improved by pretraining with previously acquired data; preferably with the same k-space pattern and receive coil. This may enable reducing the number of epochs and thus the computation time, while avoiding bias in reconstructed images due to the training dataset.

To conclude, APIR-Net provides a promising alternative to the conventional parallel imaging methods, and results in improved image quality especially for low SNR acquisitions.

References

1. Akçakaya, M., Moeller, S., Weingärtner, S., Uğurbil, K.: Scan-specific robust artificial-neural-networks for k-space interpolation (RAKI) reconstruction: database-free deep learning for fast imaging. Magn. Reson. Med. **81**(1), 439–453 (2019)

2. Griswold, M.A., et al.: Generalized autocalibrating partially parallel acquisitions (GRAPPA). Magn. Reson. Med. **47**(6), 1202–1210 (2002)
3. Hammernik, K., et al.: Learning a variational network for reconstruction of accelerated MRI data. Magn. Reson. Med. **79**(6), 3055–3071 (2018)
4. Lønning, K., Putzky, P., Sonke, J.J., Reneman, L., Caan, M.W., Welling, M.: Recurrent inference machines for reconstructing heterogeneous MRI data. Med. Image Anal. **53**, 64–78 (2019)
5. Pruessmann, K.P., Weiger, M., Scheidegger, M.B., Boesiger, P.: SENSE: sensitivity encoding for fast MRI. Magn. Reson. Med. **42**(5), 952–962 (1999)
6. Robson, P.M., Grant, A.K., Madhuranthakam, A.J., Lattanzi, R., Sodickson, D.K., McKenzie, C.A.: Comprehensive quantification of signal-to-noise ratio and g-factor for image-based and k-space-based parallel imaging reconstructions. Magn. Reson. Med. **60**(4), 895–907 (2008)
7. Uecker, M., et al.: ESPIRiT–an eigenvalue approach to autocalibrating parallel MRI: where SENSE meets GRAPPA. Magn. Reson. Med. **71**(3), 990–1001 (2014)
8. Uecker, M., et al.: Berkeley advanced reconstruction toolbox. Proc. Int. Soc. Magn. Reson. Med. **23**, 2486 (2015)
9. Wang, S., et al.: Accelerating magnetic resonance imaging via deep learning. In: 2016 IEEE International Symposium Biomedical Imaging (ISBI), pp. 514–517 (2016)
10. Zhu, B., Liu, J.Z., Cauley, S.F., Rosen, B.R., Rosen, M.S.: Image reconstruction by domain-transform manifold learning. Nature **555**(7697), 487 (2018)

Accelerated MRI Reconstruction with Dual-Domain Generative Adversarial Network

Guanhua Wang[1], Enhao Gong[2,3], Suchandrima Banerjee[4], John Pauly[3], and Greg Zaharchuk[3(✉)]

[1] University of Michigan, Ann Arbor, MI, USA
guanhuaw@umich.edu
[2] Subtle Medical, Inc., Menlo Park, CA, USA
enhao@subtlemedical.com
[3] Stanford University, Stanford, CA, USA
{pauly,gregz}@stanford.edu
[4] GE Healthcare, Chicago, IL, USA
Suchandrima.Banerjee@ge.com

Abstract. Fast reconstruction of under-sampled acquisitions has always been a central issue in MRI reconstruction. Recently years has seen multiple studies using deep learning as a de-aliasing framework to restore the aliased image. However, restoration of fine details is still problematic, especially when dealing with noisy image datasets. Sparked by the Fourier transform relationship, this work proposed and tested a new hypothesis: can regularization be directly added in the frequency domain to correct the high-frequency imperfection? To achieve this, discriminative networks are applied in both the image domain and the frequency domain (so-called dual-domain GAN). Evaluation on multiple datasets proved that the dual-domain GAN approach is an effective way to improve the quality of accelerated MR reconstruction.

Keywords: Accelerated MRI reconstruction · Generative Adversarial Network · Frequency constraint

1 Introduction

MRI is a widely used imaging modality in the clinical setting. For most MRI applications, reducing the scanning time has multiple advantages, including reducing motion artifacts, maximizing throughput and alleviating patients' discomfort. However, aggressively reducing the time under the Nyquist sampling rate can introduce aliasing artifacts to the reconstructed image. Traditional de-aliasing methods are mainly based on parallel imaging [5,17,23] and compressed sensing technologies [4,10,12]. The former type of methods utilizes redundant information from phased-array coils. However, it requires specialized hardware and the acceleration rate is generally restricted to 4. The latter method exploited the sparsity of acquisition in different transformation fields, and it can be also combined with parallel imaging to further accelerate the acquisition [3,15,16].

© Springer Nature Switzerland AG 2019
F. Knoll et al. (Eds.): MLMIR 2019, LNCS 11905, pp. 47–57, 2019.
https://doi.org/10.1007/978-3-030-33843-5_5

To further accelerate reconstruction and improve image quality, in recent years deep learning has been applied to MR reconstruction to solve the ill-conditioned problem posed by aggressively under-sampled measurements. From historical information, deep networks learn the mapping between different transform domains, including k-space to image [2,29], k-space to k-space [6], under-sampled image to fully sampled image [7,14,20,25]. For image domain-based deep learning reconstruction frameworks, they mainly embody an encoder-decoder structure like U-Net [18] and loss functions like norm loss or perceptual loss [9,20]. The idea of generative adversarial networks (GANs) has also been taken advantage of [14,20]. Currently, one major issue of DL-based reconstruction is the recovery of fine details, which is crucial for the diagnosis of multiple kinds of lesions. Recently, there are works exploring the possibility of adding a constraint in the frequency domain to solve this problem [27]. To be specific, a norm loss is calculated between the generated k-space and fully sampled k-space. However, directly adding a regularization (like L1/L2 norm loss) can be sub-optimal for the following reasons. First, the high-frequency part of k-space we care about is often noisy, making L1/L2 loss less effective. Second, Fourier transform is a linear transformation. Frequency domain L1/L2 loss do not have a clear perceptual effect different than the image domain L1/L2 loss. In this work, it is hypothesized that the GAN approach can be more suitable, since energy distribution in k-space always follows a certain pattern, especially for a specific sequence, which is relatively easy to be learned by GANs. By enforcing the realistic energy distribution, the generator may better restore the high-frequency imperfection. In the meantime, the image-domain discriminator is remained to help generate sharp and visually favorable images [11].

To test this hypothesis, multiple datasets on both brain and knee were employed, with different types of retrospective sampling patterns. Quantitative comparison with fully sampled acquisitions was performed to evaluate the model's performance compared with image domain networks.

2   Methods

2.1   Accelerated MRI with Deep Generative Model

Similar to previous works, the model aims to restore images using the information learned from historical data. To be specific, the input of the generative network is aliased images directly reconstructed from the under-sampled data. In the training process, the fully-sampled images are deemed as the ground truth, and input images are synthesized retrospectively from original acquisitions. Different loss functions measuring the distance between the aliased image and the fully-sampled one are calculated and back-propagated to train the network.

2.2   Network Architecture

The network's structure is detailed in Figs. 1 and 2, where k stands for the kernel size. s stands for the stride size, and p stands for the padding size. For the

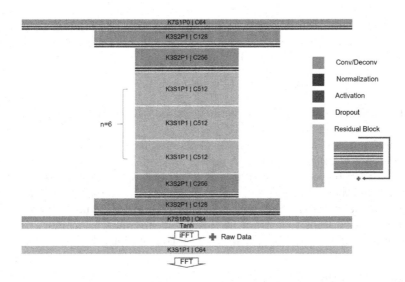

Fig. 1. The architecture of the generator.

generator, the backbone inherits the structure of ResNet [11], which has been proven to be efficient in MR reconstruction [14]. The number of Res-blocks is set the 6 to balance the model's representation power and size. Following the encoder-decoder network, an affine k-space projection layer is added, in which the real acquired k-space and the un-acquired part (which is 1-sampling mask) of the generated k-space were linearly combined. Then the combined k-space data go through several convolutional blocks before being Fourier transformed into the final image, to avoid artifact of directly combining k-space together.

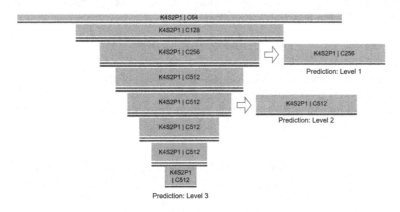

Fig. 2. The architecture of the frequency domain discriminator.

Previous works prove that combine the generated data and original acquisition can add a soft data-consistent constraint on the network, which may accelerate and stabilize the training [2,14,28].

$$k = conv(\alpha \cdot (1 - m) \cdot k_g + (1 - \alpha) \cdot m \cdot k_r) \tag{1}$$

m is the binary sampling mask, k_r is the acquired data, and k_g is the generated data. α is usually set to 0.95.

In the frequency domain, a multi-level discriminator in introduced, meaning the predictions of different output layers are all used. The design is based on the hypothesis that presentation of different levels of k-space can represent different levels of properties of k-space, and exert different effects on the generator's training. In the image domain, with trials it is found that multi-level discriminator formulation leads to a minor improvement. To minimize memory usage, a patch-based discriminator [29] consisting of 6 convolutional layers is used.

2.3   Loss Function

The loss function consists of three components: content loss, perceptual loss, and adversarial loss. Content loss measures the distance between the generated image and the ground truth in different norm spaces. In this setting, we used two sub-types, including L1/L2 norm loss and SSIM loss [8]. The L1 norm and L2 norm are mixed to balance the sharpness and stability. SSIM loss is also blended to generate more visually favorable images.

$$\min_{\theta_G} E_{x,y} \left[\|G(x) - y\|_1 + \lambda_1 \|G(x) - y\|_2 + \lambda_2 SSIM\{G(x), y\} \right] \tag{2}$$

In our implementation, λ_1 is set to 10 and λ_2 is set to 0.025.

Perceptual loss is introduced as well to restore the details better and improve the visual effect. A VGG-19 net pre-trained on ImageNet and fine-tuned on HCP dataset acted as the feature extractor [21].

$$\min_{\theta_G} E_{x,y} \left[\|\phi_j(G(x)) - \phi_j(y)\|_2 \right] \tag{3}$$

ϕ_j means the ith layer of the VGG net. The weight of the perceptual loss may vary based on the layer selected.

Adversarial loss in both the image domain and frequency domain is formulated as the LSGAN approach [13] to stabilize training.

$$\min_{\theta_G} \eta_1 E_x \sum_{i=1}^{3} \left[\gamma_i \left(1 - D_{k\_i}(FFT(G(x)))\right)^2 \right] + \eta_2 E_x \left[(1 - D_I(G(x)))^2 \right] \tag{4}$$

$$\eta_1 \beta \min_{\theta_{D_k}} E_x \sum_{i=1}^{3} \left[\gamma_i D_{k\_i}(FFT(G(x)))^2 \right] + E_y \sum_{i=1}^{3} \left[\gamma_i \left(D_{k\_i}(FFT(y)) - 1\right)^2 \right] \tag{5}$$

$$\eta_2 \beta \min_{\theta_{D_I}} E_x \left[D_I(G(x))^2 \right] + E_y \left[(D_I(y) - 1)^2 \right] \tag{6}$$

Equation 4 is the adversarial loss for the generator, Eq. 5 is for the k-space discriminator, and Eq. 6 is for the image domain discriminator. In this setting, η_1 and η_2 are usually set to 0.005. β is set to 0.25 to balance the training of the generator and the discriminator. γ is set to $[0.01, 0.1, 1]$.

Parameters were optimized via Adam optimizer. The initial learning rate is set to 1e-4 and linearly decay to 0 after 200 epochs. β parameter of Adam optimizer are set to $[0.5, 0.999]$.

For purposes of comparison, we also reimplemented other setting of loss functions: only content and perceptual loss (only generator, Gen.); content loss, perceptual loss and image domain adversarial loss (I GAN); content loss, perceptual loss and frequency domain adversarial loss (k GAN); content loss, perceptual loss and dual-domain adversarial loss (I+k GAN). The weight of the same type of loss function was tuned to its own best and remained the same across different networks to control variables.

2.4 Sampling Patterns

To testify the generalizability of the model, two popular Cartesian sampling masks, including one direction down-sampling (1D) and variable density passiondisk sampling with radial view-ordering (VDRad) [1] are adopted. For both styles, the central k-space is fully sampled to acquire more low-frequency structural information. The under-sampling rate (R) was set to 4 for both strategies.

2.5 Datasets

To evaluate the generalizability of this model, three datasets were used (Table 1). The first dataset consists of real-valued T1w and T2w structural brain imaging from the HCP project [24]. The data is acquired with MPRAGE and T2-SPACE sequences. 100 cases were included. The second dataset is a private structural brain dataset. The complex-valued images were acquired with a multi-echo saturation recovery sequence (MDME) [22,26]. All 8 echoes of the sequence were included. The third dataset is a complex-valued 3D FSE knee dataset, opensourced by Stanford and Berkley [19]. 20 cases were included.

The real part and imaginary part of the complex-valued images were decomposed into two input channels. For real-valued images, the phase was assumed to be constant. Pre-processing included zero-padding and normalization. During training, flipping and shifting were applied for purposes of data augmentation.

Table 1. Datasets

Dataset	FOV (mm)	Num. of slices	Thickness (mm)	Resolution	Sequence
HCP	224 * 224	256	N/A	0.7 (iso)	3D MPRAGE (T1w), T2-SPACE (T2w)
Knee	160 * 153.6	320	0.5	0.6 * 0.5	3D CUBE
MDME	240 * 216	31	4/4.5	0.83	Multi-dynamic, Multi-echo

2.6 Evaluation Metrics

Metrics, including peak signal-to-noise ratio (PSNR) and structural similarity metrics (SSIM), were used to evaluate the similarity with fully sampled acquisitions.

3 Results

The quantitative metrics are shown in Table 2. Compared with previous models, including ResNet (only generator, Gen.) and image-domain GAN (I GAN), most of the items were improved. Adding k-space adversarial loss only (k GAN) can also improve the performance compared with the pure generator.

Table 2. Quantitative metrics

Dataset	Sampling pattern	SSIM				PSNR(dB)			
		Gen.	I GAN	k GAN	I+k GAN	Gen.	I GAN	k GAN	I+k GAN
HCP (n = 1000)	1D	0.901	0.902	0.889	**0.909**	15.756	15.692	15.645	**16.475**
	VDRad	0.943	0.946	0.944	**0.948**	17.403	17.477	17.462	**17.509**
Knee (n = 640)	1D	0.860	**0.865**	0.861	0.864	13.606	13.784	13.790	**13.795**
	VDRad	0.872	0.873	0.872	**0.873**	14.675	14.704	14.682	**14.723**
MDME (n = 1000)	1D	0.948	0.950	0.957	**0.958**	16.682	16.816	16.996	**17.065**
	VDRad	0.974	0.974	0.974	**0.975**	18.480	18.445	18.501	**18.515**

Examples of three datasets are shown in Figs. 3, 4 and 5. Compared with previous models, less distortion, better sharpness, and contrast is achieved.

An example of restored k-space was displayed in Fig. 6. The un-acquired part and high-frequency part are better recovered than image-domain models.

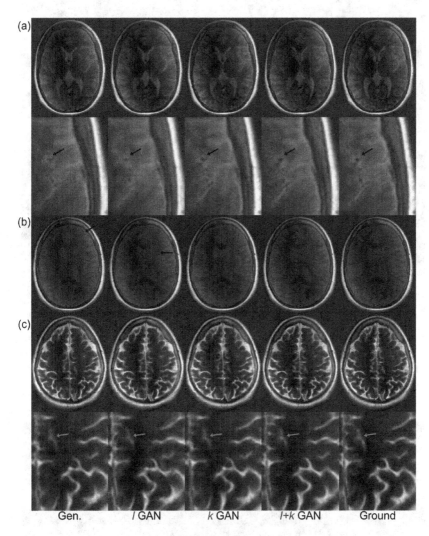

Fig. 3. Comparison of reconstruction artifacts on MDME dataset. (a), (b) and (c) are from three different echoes. In (a), the proposed method (I+k GAN) achieve the best sharpness and tissue contrast. K-space GAN also improves tissue contrast. However, the restoration of the putamen is still not perfect for all reconstruction methods. In (b), the generator only (Gen.) leads to strong blurring. While image-domain GAN improved the sharpness, strong aliasing artifacts remain. I+k GAN ensures both artifact suppression and fidelity compared to previous models. In (c), the proposed model leads to less distortion and increased sharpness of structures like sulcus.

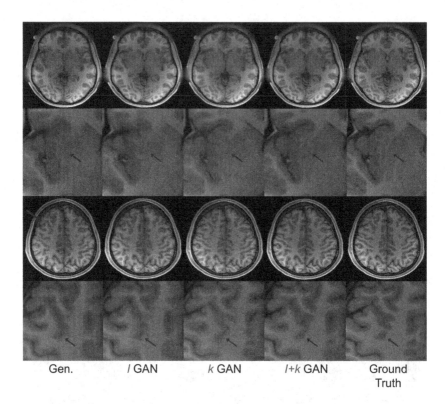

Fig. 4. Comparison of reconstruction artifacts of MDME dataset. Two T1w examples are shown. Still, the dual-domain GAN leads to the least distortion.

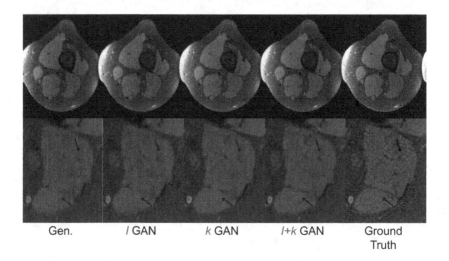

Fig. 5. Comparison of reconstruction effect of 3D knee dataset.

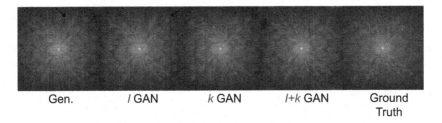

Gen.   *l* GAN   *k* GAN   *l+k* GAN   Ground
                                                                    Truth

Fig. 6. An example of logarithmic k-space from HCP dataset. The high-frequency part of k-space reconstructed by the dual-domain GAN is filled better, which corresponds to better detail restoration. The sub-sampled part of k-space is also better recovered, leading to less aliasing artifacts.

4   Discussion and Conclusions

This study validates the availability of frequency adversarial loss for DL-based MR reconstruction. Quantitative comparison with full acquisitions demonstrates that the additional frequency constraint leads to improved similarity. Visual inspection also demonstrates less geometrical distortion and better tissue contrast brought by the dual-domain GAN.

Similar to other GAN-based image restoration tasks, the hyper-parameter setting, especially weights of different loss functions, is still a tricky problem. It is supposed that the norm loss (L1/L2 norm loss) recover the main structure and adversarial loss contribute to the reconstruction of fine details. Therefore, the value of norm loss should be larger than the adversarial loss to recover the main anatomical structure better and avoid hallucination, and the weights were set to meet this standard. As a test of generalizability, the same parameter setting was used across all datasets.

Compared to the other two datasets, knee dataset is noisier, and consequently are more challenging for reconstruction algorithms. Though the visual effect is not obviously improved, the contrast between different tissues is still enhanced. The quantitative comparison also confirms the improvement. With the variable density radial sampling mask, the difference between different reconstruction networks is also not very obvious. The reason may lie in that the sampling pattern mainly introduces local blurring, which is easier to be resolved than the global blurring brought by the one-direction Cartesian sampling mask. At the same time, the better representation power and larger receptive field of deep networks are not fully utilized to unwrap the local aliasing artifacts, and simple L1/L2 norm loss is enough to train the network.

This technique can be seamlessly combined with many other DL-based reconstruction studies, either image-based or kspace-based, since it only adds a differentiable regularization to the generative model. Additionally, modern deep learning frameworks like *PyTorch* and *TensorFlow* are adding the support of FFT now; therefore, the differentiable FFT is also not an implementation burden. The main practical concern of this technique is the larger computation memory usage in the training stage, which would be alleviated along with rapid development of GPUs and ASICs.

For further studies, the multi-reader study should be conducted to evaluate clinical performance. To testify the generalizability of this model, more datasets, including more organs, scanning protocols, and pathological indications should also be applied.

References

1. Cheng, J.Y., Zhang, T., Alley, M.T., Lustig, M., Vasanawala, S.S., Pauly, J.M.: Variable-density radial view-ordering and sampling for time-optimized 3D Cartesian imaging. In: Proceedings of the ISMRM Workshop on Data Sampling and Image Reconstruction (2013)
2. Eo, T., Jun, Y., Kim, T., Jang, J., Lee, H.J., Hwang, D.: KIKI-net: cross-domain convolutional neural networks for reconstructing undersampled magnetic resonance images. Magn. Reson. Med. **80**, 2188–2201 (2018)
3. Feng, L., et al.: Golden-angle radial sparse parallel MRI: combination of compressed sensing, parallel imaging, and golden-angle radial sampling for fast and flexible dynamic volumetric MRI. Magn. Reson. Med. **72**(3), 707–717 (2014)
4. Gamper, U., Boesiger, P., Kozerke, S.: Compressed sensing in dynamic MRI. Magn. Reson. Med. **59**(2), 365–373 (2008). An Official Journal of the International Society for Magnetic Resonance in Medicine
5. Griswold, M.A., et al.: Generalized autocalibrating partially parallel acquisitions (GRAPPA). Magn. Reson. Med. **47**(6), 1202–1210 (2002). An Official Journal of the International Society for Magnetic Resonance in Medicine
6. Han, Y., Ye, J.C.: k-space deep learning for accelerated MRI. arXiv preprint arXiv:1805.03779 (2018)
7. He, K., Zhang, X., Ren, S., Sun, J.: Deep residual learning for image recognition. In: Proceedings of the IEEE Conference on Computer Vision and Pattern Recognition, pp. 770–778 (2016)
8. Hore, A., Ziou, D.: Image quality metrics: PSNR vs. SSIM. In: 2010 20th International Conference on Pattern Recognition (ICPR), pp. 2366–2369. IEEE (2010)
9. Johnson, J., Alahi, A., Fei-Fei, L.: Perceptual losses for real-time style transfer and super-resolution. In: Leibe, B., Matas, J., Sebe, N., Welling, M. (eds.) ECCV 2016. LNCS, vol. 9906, pp. 694–711. Springer, Cham (2016). https://doi.org/10.1007/978-3-319-46475-6_43
10. Jung, H., Sung, K., Nayak, K.S., Kim, E.Y., Ye, J.C.: k-t FOCUSS: a general compressed sensing framework for high resolution dynamic MRI. Magn. Reson. Med. **61**(1), 103–116 (2009). An Official Journal of the International Society for Magnetic Resonance in Medicine
11. Ledig, C., et al.: Photo-realistic single image super-resolution using a generative adversarial network. In: CVPR, vol. 2, p. 4 (2017)
12. Lustig, M., Donoho, D., Pauly, J.M.: Sparse MRI: the application of compressed sensing for rapid MR imaging. Magn. Reson. Med. **58**(6), 1182–1195 (2007). An Official Journal of the International Society for Magnetic Resonance in Medicine
13. Mao, X., Li, Q., Xie, H., Lau, R.Y., Wang, Z., Paul Smolley, S.: Least squares generative adversarial networks. In: Proceedings of the IEEE International Conference on Computer Vision, pp. 2794–2802 (2017)

14. Mardani, M., et al.: Deep generative adversarial neural networks for compressive sensing MRI. IEEE Trans. Med. Imaging **38**(1), 167–179 (2019)
15. Murphy, M., Alley, M., Demmel, J., Keutzer, K., Vasanawala, S., Lustig, M.: Fast ℓ_1-SPIRiT compressed sensing parallel imaging MRI: scalable parallel implementation and clinically feasible runtime. IEEE Trans. Med. Imaging **31**(6), 1250–1262 (2012)
16. Otazo, R., Kim, D., Axel, L., Sodickson, D.K.: Combination of compressed sensing and parallel imaging for highly accelerated first-pass cardiac perfusion MRI. Magn. Reson. Med. **64**(3), 767–776 (2010)
17. Pruessmann, K.P., Weiger, M., Scheidegger, M.B., Boesiger, P.: SENSE: sensitivity encoding for fast MRI. Magn. Reson. Med. **42**(5), 952–962 (1999)
18. Ronneberger, O., Fischer, P., Brox, T.: U-Net: convolutional networks for biomedical image segmentation. In: Navab, N., Hornegger, J., Wells, W.M., Frangi, A.F. (eds.) MICCAI 2015. LNCS, vol. 9351, pp. 234–241. Springer, Cham (2015). https://doi.org/10.1007/978-3-319-24574-4_28
19. Sawyer, A.M., et al.: Creation of fully sampled MR data repository for compressed sensing of the knee (2013)
20. Seitzer, M., et al.: Adversarial and perceptual refinement for compressed sensing MRI reconstruction. In: Frangi, A.F., Schnabel, J.A., Davatzikos, C., Alberola-López, C., Fichtinger, G. (eds.) MICCAI 2018. LNCS, vol. 11070, pp. 232–240. Springer, Cham (2018). https://doi.org/10.1007/978-3-030-00928-1_27
21. Simonyan, K., Zisserman, A.: Very deep convolutional networks for large-scale image recognition. arXiv preprint arXiv:1409.1556 (2014)
22. Tanenbaum, L.N., et al.: Synthetic MRI for clinical neuroimaging: results of the magnetic resonance image compilation (MAGiC) prospective, multicenter, multi-reader trial. Am. J. Neuroradiol. **38**, 1103–1110 (2017)
23. Uecker, M., et al.: ESPIRiT-an eigenvalue approach to autocalibrating parallel MRI: where SENSE meets GRAPPA. Magn. Reson. Med. **71**(3), 990–1001 (2014)
24. Van Essen, D.C., et al.: The WU-Minn human connectome project: an overview. Neuroimage **80**, 62–79 (2013)
25. Wang, S., et al.: Accelerating magnetic resonance imaging via deep learning. In: 2016 IEEE 13th International Symposium on Biomedical Imaging (ISBI), pp. 514–517. IEEE (2016)
26. Warntjes, J., Leinhard, O.D., West, J., Lundberg, P.: Rapid magnetic resonance quantification on the brain: optimization for clinical usage. Magn. Reson. Med. **60**(2), 320–329 (2008). An Official Journal of the International Society for Magnetic Resonance in Medicine
27. Yang, G., et al.: DAGAN: deep de-aliasing generative adversarial networks for fast compressed sensing MRI reconstruction. IEEE Trans. Med. Imaging **37**(6), 1310–1321 (2017)
28. Zhang, K., Zuo, W., Chen, Y., Meng, D., Zhang, L.: Beyond a Gaussian denoiser: residual learning of deep CNN for image denoising. IEEE Trans. Image Process. **26**(7), 3142–3155 (2017)
29. Zhu, B., Liu, J.Z., Cauley, S.F., Rosen, B.R., Rosen, M.S.: Image reconstruction by domain-transform manifold learning. Nature **555**(7697), 487 (2018)

Deep Learning for Low-Field to High-Field MR: Image Quality Transfer with Probabilistic Decimation Simulator

Hongxiang Lin[1](✉), Matteo Figini[1], Ryutaro Tanno[1,2], Stefano B. Blumberg[1], Enrico Kaden[1], Godwin Ogbole[3], Biobele J. Brown[4], Felice D'Arco[5], David W. Carmichael[6,7], Ikeoluwa Lagunju[4], Helen J. Cross[5,6], Delmiro Fernandez-Reyes[1,4], and Daniel C. Alexander[1]

[1] Centre for Medical Image Computing and Department of Computer Science, University College London, London, UK
harry.lin@ucl.ac.uk
[2] Machine Intelligence and Perception Group, Microsoft Research Cambridge, Cambridge, UK
[3] Department of Radiology, College of Medicine, University of Ibadan, Ibadan, Nigeria
[4] Department of Paediatrics, College of Medicine, University of Ibadan, Ibadan, Nigeria
[5] Great Ormond Street Hospital for Children, London, UK
[6] UCL Great Ormond Street Institute of Child Health, London, UK
[7] Department of Biomedical Engineering, King's College London, London, UK

Abstract. MR images scanned at low magnetic field (<1T) have lower resolution in the slice direction and lower contrast, due to a relatively small signal-to-noise ratio (SNR) than those from high field (typically 1.5T and 3T). We adapt the recent idea of Image Quality Transfer (IQT) to enhance very low-field structural images aiming to estimate the resolution, spatial coverage, and contrast of high-field images. Analogous to many learning-based image enhancement techniques, IQT generates training data from high-field scans alone by simulating low-field images through a pre-defined decimation model. However, the ground truth decimation model is not well-known in practice, and lack of its specification can bias the trained model, aggravating performance on the real low-field scans. In this paper we propose a probabilistic decimation simulator to improve robustness of model training. It is used to generate and augment various low-field images whose parameters are random variables and sampled from an empirical distribution related to tissue-specific SNR on a 0.36T scanner. The probabilistic decimation simulator is model-agnostic, that is, it can be used with any super-resolution networks. Furthermore we propose a variant of U-Net architecture to improve its learning performance. We show promising qualitative results from clinical low-field images confirming the strong efficacy of IQT in an important new application area: epilepsy diagnosis in sub-Saharan Africa where only low-field scanners are normally available.

© Springer Nature Switzerland AG 2019
F. Knoll et al. (Eds.): MLMIR 2019, LNCS 11905, pp. 58–70, 2019.
https://doi.org/10.1007/978-3-030-33843-5_6

1   Introduction

Magnetic Resonance Imaging (MRI) is now ubiquitous in neurology with a strong trend towards the use of high-field scanners, with 1.5T and 3T being the current clinical standard. However, low-field MRI scanners, less than 1T, are still common in low and middle income countries (LMICs), due to limited funds and frequent power outages. Low-field scanners suffer from lower signal-to-noise ratio (SNR) than high field at equivalent spatial resolution. To counteract the SNR reduction, practitioners commonly acquire images with non-adjacent thick slices to reduce the acquisition time and cross-talk artifacts in brain MRI scenario [1]. This leads to resolution reduction in the slice direction compared with the in-plane resolution and a loss of information due to gaps between slices; see Fig. 1(a–b). Moreover, the contrast between grey matter (GM) and white matter (WM) may be worse than in high field even at equivalent SNR and spatial resolution as illustrated in Fig. 1(c–d).

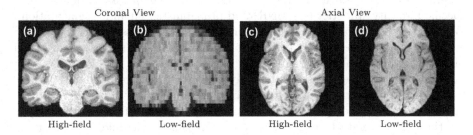

Fig. 1. High-field vs low-field MR scans: (a–b) Resolution change on coronal plane; (c–d) Contrast change on axial plane. Data sources: (a, c) 3T MRI from Human Connectome Project [2]; (b, d) 0.36T MRI acquired from University College Hospital, Ibadan.

In this study, we aim to learn an image-translation mapping from low field to high field to perform super-resolution and contrast enhancement. In the literature, mathematical models have been proposed to describe the variation of MRI signal with the magnetic field [3,4], but such models are simplistic and do not include all effects on the final images, such as variability in the acquisition process. Furthermore, the reconstruction of missing information between the acquired slices is severely ill-posed, which hinders the practical capability of producing high-field like images. Several approaches in the literature aim to solve related problems. Bahrami et al. [5] proposed a multi-level Canonical Correlation Analysis for estimating 7T from 3T images using paired training data. Wolterink et al. [6] used the idea of cycle consistency to leverage the abundance of unpaired training sets and learn to synthesise CT from MRI. This approach is, however, known to be susceptible to hallucinations and may introduce spurious features in the output images [7].

Image Quality Transfer (IQT) is a machine learning framework used to enhance low-quality clinical data to the abundant neurological information in high-quality images. Most implementations of IQT simulate low quality data from high quality providing matched-paired for training. In [8–11] for instance, the corresponding low-field data are synthesised by downsampling and matching voxel-wise intensities coming from prior or empirical knowledge about actual low-field data. However, the trained model strongly depends on the accuracy of low-field synthesis. To improve model generalisability, the prediction of a trained model should be built on unseen test data with less dependency of simulation.

In this paper, we build on the IQT framework to construct a mapping that estimates high-field images from the matched low-field inputs. The paired data, particularly in large numbers, are hard to acquire in one area due to the rare availability of high-field scanners in LMICs and low-field scanners in high income countries (HICs). Our key technical contribution is to propose a probabilistic decimation (downsampling) model to improve robustness of IQT training and to enhance images from low-field scanners. More specifically, low-field data generation comes from a probabilistic model which comprises random tissue-specific intensity statistics (e.g. SNR) and probabilistic semantic segmentation. We assume that an *a priori* distribution related to the tissue-specific SNR is available. The segmentation mask estimated by Statistical Parametric Mapping [12] is also probabilistic in terms of the tissue type. Therefore for one high-field subject, we can simultaneously generate the corresponding multiple low-field data and form the paired training data, a novel way of performing data augmentation. We then learn the low-field-to-high-field transformation by adapting the U-Net architecture [13] with a super-resolution module, a "bottleneck block", extending its depth to enable it to capture more global features of image contrast.

2   Methods

2.1   Formulation

Let a 3D low-field input patch x of size $w \times h \times d$ be corrupted by smoothing, low contrast, and random noise. It is randomly cropped from the original low-field MR volume denoted by X. Our aim is to reconstruct the sub-voxel information in the slice thickness direction and to attain the high SNR and contrast transferring to the corresponding high-field output patch y of size $w \times h \times kd$, where k is an upsampling rate. Then we assemble all output patches into a high-field MR volume denoted by Y. The relationship between x and y is modelled by a degradation process of image quality, described by a function S such that

$$x = S(y, \boldsymbol{\alpha}) + \epsilon, \tag{1}$$

where $\boldsymbol{\alpha}$ denotes a vector of SNR components corresponding to prior knowledge of WM and GM in the low-field input volume, i.e. $\boldsymbol{\alpha} = (SNR_X^{WM}, SNR_X^{GM})$. It is randomly sampled from the Gaussian distribution $\mathcal{N}(\boldsymbol{\mu}, \Sigma)$ where $\boldsymbol{\mu}$ is a mean vector and Σ is a covariance matrix. ϵ denoting background noise has a Gaussian distribution $\mathcal{N}(0, \sigma_{BG}^2)$. Section 2.2 will specify the formulation and algorithm for modelling S. We then employ deep learning, specifically a convolutional neural network, to estimate the inverse mapping $S^\dagger$.

We use a given M-paired training set $\mathcal{T}_M = \{(x_i, y_i)\}_{i=1}^M$ with a fixed α to train our convolutional neural networks over all sampled patches from all MR volumes. We optimise the network parameters θ by minimising the average of the pixel-wise mean squared error (MSE) denoted by $\| \cdot \|_2^2$ over all training sets:

$$\theta^* = \arg\min_\theta \frac{1}{M} \sum_{i=1}^M \|S_\theta^\dagger(x_i) - y_i\|_2^2. \tag{2}$$

2.2   Probabilistic Decimation Simulator

Equation (1) enables us to produce additional training data by randomly sampling the coefficient α from an *a priori* distribution, forming the so-called probabilistic decimation simulator. It translates the voxel-wise low-field SNRs, related to the sampled α and the tissue category, to the high-field image and down-samples with a factor of k. We use this simulator to generate N low-field patches for each high-field patch y_i and form a new training set $\mathcal{T}_{M,N} = \{(x_{ij}, y_i) | i = 1, \cdots, M, j = 1, \cdots, N\}$. Henceforth, the new model is trained on the augmented set $\mathcal{T}_{M,N}$ with the following expression:

$$\theta^* = \arg\min_\theta \frac{1}{MN} \sum_{i=1}^M \sum_{j=1}^N \|S_\theta^\dagger(x_{ij}) - y_i\|_2^2. \tag{3}$$

We develop Algorithm 1 for implementing the probabilistic decimation simulator for neural images. We transform high-field images $Y(\mathbf{v})$ to synthetic low-field images denoted by $\hat{X}(\mathbf{v})$ for any voxel coordinate $\mathbf{v}$ by adapting the SNR in WM and GM to the values obtained in our reference low-field dataset. We assume that SNRs of WM and GM have a 2D Gaussian distribution and the background noise in the low-field or the high-field images has a 1D Gaussian distribution with a zero mean and a standard deviation of σ_X or σ_Y, respectively. We also assume $\sigma_X \gg \sigma_Y$ since the random noise in high field is negligible. The simulation procedure starts with the skull-stripped $Y(\mathbf{v})$ with isotropic voxels of length e_z. We then down-sample along the slice thickness direction (vertical, or z-direction). A $1D$ Gaussian filter $h_\sigma(z) = \frac{1}{\sigma\sqrt{2\pi}} e^{-z^2/(2\sigma^2)}$ is applied to the high-field images along the z-direction, where the σ is linked to a full-width at half maximum (FWHM): FWHM $= 2\sqrt{2\ln 2}\sigma$. The FWHM of the Gaussian filter is set to the slice thickness, or in terms of σ: $\sigma = k e_z / \sqrt{8 \ln 2}$. Then the distance between slices is set to be larger than this slice thickness, emulating the gap between slices. The slices of the original image falling in the gaps have virtually no effect on the signal in the simulated image, similar to what happens in real acquisitions. The high-field images $Y(\mathbf{v})$ are first segmented into tissue

Algorithm 1. Probabilistic Decimation Simulator for low-field Image

---

Input: high-field Images $Y(\mathbf{v})$, masks $M^j(\mathbf{v})$ for $j = WM, GM, others$, downsampling scale $k \in \mathbb{N}$, background noise levels σ_X and σ_Y, low-field SNR distribution $\mathcal{N}(\boldsymbol{\mu}, \Sigma)$.

1: $Y_{\downarrow k}(\mathbf{v}) = Y_{\downarrow k}(\tilde{v}, v') = \sum_{v''} Y(\tilde{v}, kv' - v'')h_\sigma(v'')$; $\triangleright$ Downsample on v'' component.

2: $Y_{\downarrow k}^j(\mathbf{v}) = M^j(\mathbf{v})Y_{\downarrow k}(\mathbf{v})$; $\triangleright$ Apply masks.

3: $SNR_Y^j = \sum_{\mathbf{v}} Y_{\downarrow k}^j(\mathbf{v}) / \left(\sigma_Y \sum_{\mathbf{v}} M^j(\mathbf{v}) \right)$; $\triangleright$ Compute SNRs for high field.

4: $(SNR_X^{WM}, SNR_X^{GM}) \sim \mathcal{N}(\boldsymbol{\mu}, \Sigma)$; $\triangleright$ Sample SNRs for low field.

5: $l^j = \begin{cases} SNR_X^j / SNR_Y^j, & j = WM, GM, \\ 1, & others; \end{cases}$ $\triangleright$ Evaluate ratio of image intensity.

6: $\hat{X}(\mathbf{v}) = \sum_{j \in \{WM, GM, others\}} l^j Y_{\downarrow k}^j(\mathbf{v})$; $\triangleright$ Transfer contrast.

7: $\hat{X}_\epsilon(\mathbf{v}) = \hat{X}(\mathbf{v}) + \epsilon(\mathbf{v})$ where $\epsilon(\mathbf{v}) \sim \mathcal{N}(0, \sigma_X^2)$. $\triangleright$ Add noise.

Output: Noisy synthetic low-field image $\hat{X}_\sigma(\mathbf{v})$.

---

categories $j = WM, GM, others$ (denoted by $M^j(\mathbf{v})$) using the unified segmentation algorithm in Statistical Parametric Mapping [12]. In this algorithm, the mask $M^j(\mathbf{v})$ corresponds to the probability that each voxel $\mathbf{v}$ belongs to the tissue category j. SNR of the high-field image with respect to the tissue category j is defined as:

$$SNR_Y^j = \frac{\sum_{\mathbf{v}} M^j(\mathbf{v}) Y(\mathbf{v})}{\sigma_Y \sum_{\mathbf{v}} M^j(\mathbf{v})}. \tag{4}$$

This allows us to evaluate ratios of low-field-to-high-field image intensity for both WM and GM; see Step 5. We then re-scale the high-field images with the ratios of image intensity according to tissue category, which results in the synthetic low-field images $\hat{X}(\mathbf{v})$. We finally add Gaussian white noise to $\hat{X}(\mathbf{v})$, with a standard deviation of σ_X.

2.3 Deep Learning Framework

The classical 3D isotropic U-Net [14] maps two identical-size cubes serving as input and output through the encoder-decoder framework. Each level, defined as a collection of operations in between two shape deformations, for a typical U-Net consists of several convolutional layers together with a pooling layer. The activation from each level in the encoder is concatenated to the input features to the same level in the decoder, enabling the network to integrate both local and global image features. U-Net uses the "same" zero-padding technique so that feature sizes keep invariant during convolution.

In this work, we extend the U-Net architecture into mapping input and output patches differing with up-scaling factor k in the slice direction. Considering the case of $k = 4$ illustrated in Fig. 2, this anisotropic U-Net first partially downsamples the first two dimensions until the down-scaling features become isotropic and thereafter conducts isotropic down- and up-sampling. To achieve this, we define the following two operations:

Bottleneck Block. To incorporate a super-resolution transformation into U-Net, we propose a bottleneck block used to connect corresponding levels of the

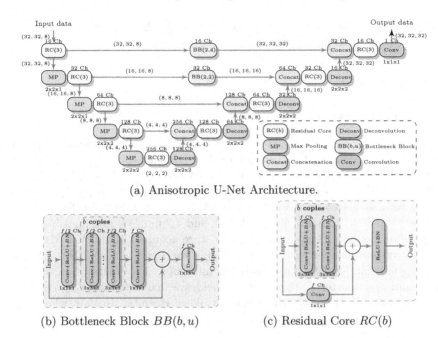

(a) Anisotropic U-Net Architecture.

(b) Bottleneck Block $BB(b, u)$ (c) Residual Core $RC(b)$

Fig. 2. (a) The diagram of anisotropic U-Net (example for the up-scaling factor of $k = 4$). The operations, (b) Bottleneck Block $BB(b, u)$ with f filters and (c) Residual Core $RC(b)$ with f filters, are detailed. The round boxes correspond to the different operations illustrated in the bottom right of (a). The number of output channels, abbreviated as "Ch", and the kernel size are denoted on top and bottom of the boxes. The arrows represent transfer of data with its corresponding shape highlighted.

contracting and expanding paths, as shown in Fig. 2(b). The design is inspired by bottleneck block in ResNet [15] and FSRCNN [16]. The bottleneck block $BB(b, u)$ has three hyperparameters: the input filter f, the number of shrinking layers b and the up-sampling scaling factor u. It shrinks half of the filters on consecutive $3 \times 3 \times 3$ convolutional layers between two endpoint convolutions with a kernel size of $1 \times 1 \times 1$. All convolution layers are activated by Rectified Linear Unit (ReLU) with Batch Normalization (BN). The skip connection enables the training of deeper networks [15]. Resolution change is efficiently carried out by a transpose convolution, or deconvolution, with the same kernel and stride of $(1, 1, u)$.

Residual Core. To have more convolutional layers on each level, the residual core that is a revision of residual element in [17] is introduced in Fig. 2(c). This is a combination of several sequential $3 \times 3 \times 3$ convolutional layers, followed by ReLU and BN layers, skip connected with an $1 \times 1 \times 1$ fully convolutional layer. Then the output is attained before ReLU and BN again. Utilizing the consecutive convolutional layers enlarges each receptive field on each level. Moreover, the appended skip connection is able to avoid the vanishing gradient problem in neural networks with gradient-based learning methods.

3   Experiments

3.1   Implementation Details

Datasets. High-resolution axial T1-weighted images were obtained from the publicly available Human Connectome Project (HCP) dataset [2], acquired on a 3T Siemens Connectome scanner with an isotropic voxel size of $0.7 \times 0.7 \times 0.7$ mm^3. To investigate sensitivity of the proposed U-Net, we trained it on two training sets with two up-scaling factors of $k = 4$ or 8. Specifically, the slice thickness/gap is 2.1 mm/0.7 mm for $k = 4$, and 4.2 mm/1.4 mm for $k = 8$. As a reference for low field, T1-weighted images were acquired on a 0.36T MagSense 360 MRI System scanner with a non-isotropic voxel size of $0.9 \times 0.9 \times 7.2$ mm^3 including 6.0 mm slice thickness and 1.2 mm gaps. The distribution of white matter and grey matter SNRs in the low field was acquired from 28 image data from children with epilepsy in University College Hospital, Ibadan, whose ages are within a range from 2 to 15 years.

IQT Pipeline. In the training stage, we randomly selected 30 subjects with skull-stripping from HCP dataset and employed them to synthesise the low-field images using Algorithm 1 based on *a priori* variable SNRs. Regarding patch extraction, we cropped the low-field patches with the step size of 8, 16, and $16/k$ along x-, y-, and z-directions, respectively. We also cropped the high-field patches with the same volume and position as the corresponding low-field patches. The low-field and high-field patch sizes were $32 \times 32 \times (32/k)$ and $32 \times 32 \times 32$, respectively. Then the patches capturing 80% background voxels were excluded from a patch library.

We examined if overfitting occurred with a validation set and judged the performance of the trained neural network with an evaluation set. We split all 30 subjects into 12, 3, and 15 for training, validation, and evaluation sets. Moreover, we investigated the image quality by calculating the peak signal-to-noise ratio (PSNR) and the structural similarity index (SSIM) [18]. We employed a two-tailed Wilcoxon signed-rank test to determine the statistical significance of the performance difference between two comparing methods.

Neural Networks. We conducted an ablation study on the proposed U-Net, denoted by ANISO U-Net(b), in the case of $b = 2$ or 3 for shrinking layers in the bottleneck block. We evaluated our networks against the 3D cubic B-spline interpolation and several existing U-Net baselines equivalently switching off the corresponding blocks, i.e. bottleneck block and residual core, in ANISO U-Net. One is an isotropic 3D U-Net (ISO U-Net) [14] implemented with 5 levels and 3 convolutional layers per level. The input of ISO U-Net is isotropically interpolated using cubic B-splines. The other one is 3D-SRU-Net [13] that up-samples each level output on the contraction path before concatenation. It contained 3

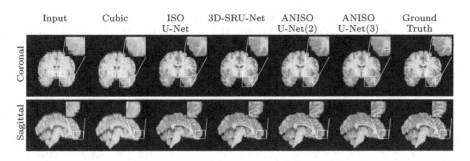

Fig. 3. Visualization of U-Net reconstructions with the up-scaling factor $k = 8$.

levels for the down-sampling scale $k = 4$ and 4 levels for $k = 8$. We unified hyper-parameters of the three U-Nets as follows. Number of filters on the first level was 16 with the number of filters doubling at each subsequent level. All U-Nets were implemented in Python using Keras library [19] with Tensorflow backend. They were calculated on a Nvidia GTX 1080 Ti GPU. Training used ADAM [20] as the optimizer with a starting learning rate of 10^{-3} and a decay of 10^{-6}. We initialized the parameters with Glorot normal initializer [21]. The batchsize was 32 and the loss function is the pixel-wise mean squared error (MSE). All the experiments started converging after about 30 epochs and we employed early stopping after 5 epochs of no improvement on the validation set.

3.2   Evaluation on Fixed SNR Data Sets

We evaluated the ability of the proposed U-Net in an ideal case where the SNR-related coefficient $\alpha = (SNR_X^{WM}, SNR_X^{GM})$ in Eq. (1) is deterministic. We fixed the SNR_X^{WM} and SNR_X^{GM} as 61 and 53, respectively, in the IQT pipeline by reconstructing images in the evaluation set at Step 4 in Algorithm 1. Table 1 shows that our model, ANISO U-Net(2), achieved the best performance in terms of the average PSNR and SSIM, and especially, significantly outperformed the others in terms of PSNR at $k = 4$ and the mean SSIM (MSSIM) at $k = 8$. The reconstruction degraded as the up-scaling factor increased. Figure 3 shows the U-Net reconstructions on coronal and sagittal planes. Qualitatively we observed clear recovery of high resolution information and enhancement of contrast. The reconstructed images from all networks nicely highlighted features visible in the ground truth images that were obscured in the low quality input. The quantitative results in Table 1 show little difference among the U-Net outputs but they might not be able to reflect subtle qualitative differences. The zoomed patches in Fig. 3 highlight differences more clearly and we believe ANISO U-Net(2) approximates the ground truth most closely and with the least artefacts as shown in the ANISO U-Net(3) result of Fig. 3. Delicately selecting hyper-parameters can avoid overfitting, and hence can mitigate the artifacts.

Table 1. The performance of the proposed model on up-scaling factors $k = 4$ or 8. The mean and standard deviation of PSNR and the mean SSIM (MSSIM) are calculated over 15 evaluation subjects. For each case, we show the best performance over an ensemble of 5 trained models. Bold font denotes the best mean or standard deviation. The asterisk * denotes p-value < 0.01 compared with the rest methods.

Method	$k = 4$		$k = 8$	
	PSNR (dB)	MSSIM	PSNR (dB)	MSSIM
Cubic B-spline	20.689 ± 2.540*	0.692 ± 0.0384*	18.974 ± 2.535*	0.567 ± 0.0471*
ISO U-Net	30.798 ± 2.573	0.916 ± 0.0227*	27.073 ± 2.469	0.846 ± 0.0278
3D-SRU-Net	30.764 ± 2.638	0.922 ± 0.0191	27.275 ± 2.542	0.847 ± 0.0290
ANISO U-Net(2)	$\mathbf{31.045 \pm 2.654}$*	$\mathbf{0.923 \pm 0.0197}$	$\mathbf{27.346 \pm 2.517}$	$\mathbf{0.852 \pm 0.0280}$*
ANISO U-Net(3)	30.918 ± 2.639	0.921 ± 0.0199*	27.054 ± 2.544	0.847 ± 0.0282

3.3   Evaluation on Variable SNR Data Sets

We evaluated the performance of several deep learning architectures including the proposed anisotropic U-Net with variable-SNR low-field data. SNR_X^{WM} and SNR_X^{GM} are now sampled from a two-dimensional Gaussian distribution $\mathcal{N}(\mu, \Sigma)$ where the coefficients are:

$$\mu = (64.50, 54.14) \quad \Sigma = \begin{pmatrix} 78.47, 71.50 \\ 71.50, 73.91 \end{pmatrix}.$$

The simulator shown in Algorithm 1 randomly generated N low-field input images with different SNR for the chosen 15 training subjects in the HCP data set. We trained the deep learning models on the dataset with the augmenting factor $N = 1, 2, 4$ and 8. We randomly selected 12.5% patches for training in each overlap patch library. For each neural network, an ensemble of 5 models were trained in terms of different augmented dataset.

Table 2 shows the mean and standard deviation of PSNR and MSSIM over 15 test subjects in terms of the augmented datasets and deep learning architectures. As a result, probabilistic decimation model was generally able to produce more stable reconstruction than the deterministic model if the unseen test data were also generated from the variable SNR. Both accuracy and robustness corresponding to mean and standard deviation of MSSIM improved in various degree as the number of generated low-field image samples increased, and in addition, the performances for the two methods were statistically significant in terms of $N = 8$ at $k = 8$. Regarding PSNR, the performance upgraded after augmentation but the robustness reflected by the standard deviation did not improve correspondingly. In addition, we observed that PSNR and MSSIM at $k = 4$ only slightly improve when the augmenting factor N became larger, which means the improvement of performance arising from augmentation gradually reached an upper bound.

Table 2. The performance of probabilistic decimation simulation for augmentation with a factor of N. The mean and standard deviation of PSNR and MSSIM are calculated over 15 evaluation subjects. We show the best performance over an ensemble of 5 trained models. The "const" at N samples/subject column means that the models were trained on the fixed SNR data sets as described in Sect. 3.2. Bold font denotes the best mean or standard deviation. The asterisk * denotes p-value< 0.01 compared with the other augmentation factors.

Method	N samples /subject	$k = 4$		$k = 8$	
		PSNR (dB)	MSSIM	PSNR (dB)	MSSIM
3D-SR U-Net	"const"	27.214 ± 3.030	0.871 ± 0.0332	24.687 ± 2.464	0.758 ± 0.0371
	1	$27.988 \pm \mathbf{2.445}$	0.861 ± 0.0286	23.777 ± 2.693	0.757 ± 0.0401
	2	29.453 ± 2.585	$0.901 \pm 0.0240^*$	$25.513 \pm 2.711^*$	$0.799 \pm 0.0366^*$
	4	$\mathbf{30.257} \pm 2.647$	$\mathbf{0.918} \pm \mathbf{0.0201}$	26.025 ± 2.587	$0.816 \pm 0.0339^*$
	8	29.958 ± 2.541	0.911 ± 0.0203	$\mathbf{26.391} \pm 2.621$	$\mathbf{0.832} \pm \mathbf{0.0316}^*$
ANISO U-Net(2)	"const"	27.311 ± 3.522	0.870 ± 0.0338	24.754 ± 2.367	0.769 ± 0.0341
	1	28.664 ± 2.552	0.890 ± 0.0240	$23.418 \pm \mathbf{2.322}$	$0.757 \pm 0.0345^*$
	2	$29.216 \pm \mathbf{2.308}$	0.893 ± 0.0248	$25.862 \pm 2.617^*$	0.803 ± 0.0343
	4	30.248 ± 2.565	$\mathbf{0.916} \pm \mathbf{0.0195}$	26.231 ± 2.613	0.807 ± 0.0381
	8	$\mathbf{30.344} \pm 2.421$	0.914 ± 0.0188	$\mathbf{27.053} \pm 2.398$	$\mathbf{0.843} \pm \mathbf{0.0330}^*$

3.4   Test on Patient Data

We tested our IQT approach on the data from a 10-year-old epilepsy patient who has two cortical-subcortical cystic lesions with surrounding edema on low-field T1-weighted images at the GM-WM junction of the parietal lobes. In this case, we used IQT with ANISO U-Net(2) trained on the HCP dataset with the augmenting factor $N = 1$ and the up-scaling factor of $k = 4$. Figure 4 shows the axial and coronal results enhanced from the low-field T1-weighted image of the patient. The IQT approach improved the GM-WM contrast globally, and significantly enhanced the resolution in coronal and sagittal planes. The enhanced image strongly highlights the two lesions in this patient which are very subtle on the input T1-weighted image. In this particular patient, the lesions were clearly visible on the original T2-weighted image, which validates that IQT highlights the lesions in the correct locations, as Fig. 4(c) shows. However, in general not all lesions are clearly visible on any MRI sequence, especially at low field, and Fig. 4 highlights the potential of our algorithms to reveal subtle lesions enhancing diagnosis and potentially enabling effective treatment via clear localisation.

4   Discussion and Conclusion

In this work, we present an IQT approach to enhance low-field MRIs aiming to match resolution as well as contrast of high-field images. We introduce the anisotropic U-Net characterised by a deeper hierarchy and super resolving connections between input and output layers. We propose the probabilistic decimation simulator by synthesising multiple low-field images with respect to distinct

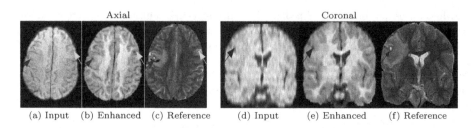

(a) Input   (b) Enhanced   (c) Reference   (d) Input   (e) Enhanced   (f) Reference

Fig. 4. The IQT prediction on the low-field epileptic patient data for (a–c) axial plane and (d–f) coronal plane. (a) and (d): Low-field T1-weighted input with cubic B-spline interpolation; (b) and (e): IQT-enhanced T1-weighted output using ANISO U-Net(2); (c) and (f): low-field T2-weighted image as a reference of ground truth. Two sub-centimeter parenchymal cystic lesions at the GM-WM junction of the parietal lobes are pointed out by the red and the yellow arrows. They are barely visible in (a) and (d) but greatly enhanced in (b) and (e). (c) and (f), not involved in the IQT experiment, verified their location in an independent acquisition.

grey-white matter SNR sampled from an *a priori* distribution. We demonstrate that the proposed method improves the robustness on the unseen test data of variable SNR at the evaluation stage. We validate our proposed U-Net on the evaluation dataset and the results potentially show generalisability to the actual clinical low-field images.

This work offers several avenues for future improvement and application. Here the metrics (MSSIM and PSNR) used for quantitative assessment reflect the performance on only synthetic images. This demonstrates efficacy, but evaluation on a sizeable data set of clinical images and clinical significance from radiologists are essential for further translation. Therefore, additional qualitative evaluation by radiologist ratings and, ultimately, demonstration of improved decision making is essential to confirm impact of the approach. Nevertheless, we believe our methods have great potential to identify subtle lesions in epilepsy and other neurological conditions and thus to improve patient outcomes in LMICs in the future.

Acknowledgements. This work was supported by EPSRC grants (EP/R014019/1, EP/R006032/1 and EP/M020533/1) and the NIHR UCLH Biomedical Research Centre. Data were provided in part by the Human Connectome Project, WU-Minn Consortium (Principal Investigators: David Van Essen and Kamil Ugurbil; 1U54MH091657) funded by NIH and Washington University. The 0.36T MRI data were acquired at the University College Hospital, Ibadan, Nigeria.

References

1. Wadghiri, Y.Z., Johnson, G., Turnbull, D.H.: Sensitivity and performance time in MRI dephasing artifact reduction methods. Magn. Reson. Med. **45**(3), 470–476 (2001)

2. Sotiropoulos, S.N., et al.: Advances in diffusion MRI acquisition and processing in the human connectome project. NeuroImage **80**, 125–143 (2013)
3. Marques, J.P., Simonis, F.F.J., Webb, A.G.: Low-field MRI: an MR physics perspective. J. Magn. Reson. Imaging **49**(6), 1528–1542 (2019)
4. Brown, R.W., Cheng, Y.-C.N., Haacke, E.M., Thompson, M.R., Venkatesan, R.: Magnetic Resonance Imaging: Physical Principles and Sequence Design, 2nd edn. Wiley, Hoboken (2014)
5. Bahrami, K., Shi, F., Zong, X., Shin, H.W., An, H., Shen, D.: Reconstruction of 7T-Like Images from 3T MRI. IEEE Trans. Med. Imaging **35**(9), 2085–2097 (2016)
6. Wolterink, J.M., Dinkla, A.M., Savenije, M.H.F., Seevinck, P.R., van den Berg, C.A.T., Išgum, I.: Deep MR to CT synthesis using unpaired data. In: Tsaftaris, S.A., Gooya, A., Frangi, A.F., Prince, J.L. (eds.) SASHIMI 2017. LNCS, vol. 10557, pp. 14–23. Springer, Cham (2017). https://doi.org/10.1007/978-3-319-68127-6_2
7. Cohen, J.P., Luck, M., Honari, S.: Distribution matching losses can hallucinate features in medical image translation. In: Frangi, A.F., Schnabel, J.A., Davatzikos, C., Alberola-López, C., Fichtinger, G. (eds.) MICCAI 2018. LNCS, vol. 11070, pp. 529–536. Springer, Cham (2018). https://doi.org/10.1007/978-3-030-00928-1_60
8. Alexander, D.C., et al.: Image quality transfer and applications in diffusion MRI. NeuroImage **152**, 283–298 (2017)
9. Tanno, R., Ghosh, A., Grussu, F., Kaden, E., Criminisi, A., Alexander, D.C.: Bayesian image quality transfer. In: Ourselin, S., Joskowicz, L., Sabuncu, M.R., Unal, G., Wells, W. (eds.) MICCAI 2016. LNCS, vol. 9901, pp. 265–273. Springer, Cham (2016). https://doi.org/10.1007/978-3-319-46723-8_31
10. Tanno, R., Worrall, D.E., Ghosh, A., Kaden, E., Sotiropoulos, S.N., Criminisi, A., Alexander, D.C.: Bayesian image quality transfer with CNNs: exploring uncertainty in dMRI super-resolution. In: Descoteaux, M., Maier-Hein, L., Franz, A., Jannin, P., Collins, D.L., Duchesne, S. (eds.) MICCAI 2017. LNCS, vol. 10433, pp. 611–619. Springer, Cham (2017). https://doi.org/10.1007/978-3-319-66182-7_70
11. Blumberg, S.B., Tanno, R., Kokkinos, I., Alexander, D.C.: Deeper image quality transfer: training low-memory neural networks for 3D images. In: Frangi, A.F., Schnabel, J.A., Davatzikos, C., Alberola-López, C., Fichtinger, G. (eds.) MICCAI 2018. LNCS, vol. 11070, pp. 118–125. Springer, Cham (2018). https://doi.org/10.1007/978-3-030-00928-1_14
12. Ashburner, J., Friston, K.J.: Unified segmentation. NeuroImage **26**, 839–851 (2005)
13. Heinrich, L., Bogovic, J.A., Saalfeld, S.: Deep learning for isotropic super-resolution from non-isotropic 3D electron microscopy. In: Descoteaux, M., Maier-Hein, L., Franz, A., Jannin, P., Collins, D.L., Duchesne, S. (eds.) MICCAI 2017. LNCS, vol. 10434, pp. 135–143. Springer, Cham (2017). https://doi.org/10.1007/978-3-319-66185-8_16
14. Çiçek, Ö., Abdulkadir, A., Lienkamp, S.S., Brox, T., Ronneberger, O.: 3D U-Net: learning dense volumetric segmentation from sparse annotation. In: Ourselin, S., Joskowicz, L., Sabuncu, M.R., Unal, G., Wells, W. (eds.) MICCAI 2016. LNCS, vol. 9901, pp. 424–432. Springer, Cham (2016). https://doi.org/10.1007/978-3-319-46723-8_49
15. He, K., Zhang, X., Ren, S., Sun, J.: Deep residual learning for image recognition. CVPR **2016**, 770–778 (2016)
16. Dong, C., Loy, C.C., Tang, X.: Accelerating the super-resolution convolutional neural network. In: Leibe, B., Matas, J., Sebe, N., Welling, M. (eds.) ECCV 2016. LNCS, vol. 9906, pp. 391–407. Springer, Cham (2016). https://doi.org/10.1007/978-3-319-46475-6_25

17. Guerrero, R., et al.: White matter hyperintensity and stroke lesion segmentation and differentiation using convolutional neural networks. NeuroImage Clin. **17**, 918–934 (2018)
18. Wang, Z., Bovik, A.C., Sheikh, H.R., Simoncelli, E.P.: Image quality assessment: from error visibility to structural similarity. IEEE Trans. Image Process. **13**(4), 600–612 (2004)
19. Chollet, F., et al.: Keras (2015). https://keras.io
20. Kingma, D.P., Ba, J.: Adam: a method for stochastic optimization. arXiv preprint arXiv:1412.6980 (2014)
21. Glorot, X., Bengio, Y.: Understanding the difficulty of training deep feedforward neural networks. In: AISTATS 2010, PMLR, vol. 9, pp. 249–256 (2010)

Joint Multi-anatomy Training of a Variational Network for Reconstruction of Accelerated Magnetic Resonance Image Acquisitions

Patricia M. Johnson[1(✉)], Matthew J. Muckley[1], Mary Bruno[1],
Erich Kobler[2], Kerstin Hammernik[2,3], Thomas Pock[2],
and Florian Knoll[1]

[1] Radiology Department, Center for Biomedical Imaging,
NYU Langone Health, New York, NY, USA
`patricia.johnson3@nyulangone.org`
[2] Institute of Computer Graphics and Vision, Graz University of Technology,
Graz, Austria
[3] Department of Computing, Imperial College London, London, UK

Abstract. Magnetic resonance imaging is a leading image modality for many clinical applications; however, a significant drawback is the lengthy data acquisition. This motivates the development of methods for reconstruction of sparsely sampled image data. One such technique is the Variational Network (VN), a machine learning method that generalizes traditional iterative reconstruction techniques, learning the regularization term from large amounts of image data. Previously, with the VN technique, reconstruction of 4-fold accelerated knee images was shown to be highly successful. In this work we extend the VN approach to applications beyond knee imaging and evaluate the classic VN and a newly developed Unet-VN in 5 different anatomical regions. We evaluate the networks trained individually for each anatomical area as well as jointly trained with data from all anatomical areas. The VN and Unet-VN were trained to reconstruct 4-fold accelerated images of knees, brains, hips, ankles and shoulders. SSIM was calculated to quantitatively evaluate the reconstructed images. Results show that the Unet-VN outperforms the classic VN, both quantitatively – in terms of structural similarity – and qualitatively. The networks jointly trained with multi-anatomy data approach the performance of the individually trained networks and offer the simplicity of a single network for a range of clinical applications which has substantial benefit for clinical translation.

Keywords: MR image reconstruction · Variational network · Machine learning

1 Introduction

The acquisition of Magnetic Resonance Image (MRI) data is an inherently slow process due to the high sampling requirements. Reconstructing images with sparser sampling has been, and continues to be, an active area of research in MRI. The major developments that have contributed to faster imaging are parallel imaging [1–3] and

© Springer Nature Switzerland AG 2019
F. Knoll et al. (Eds.): MLMIR 2019, LNCS 11905, pp. 71–79, 2019.
https://doi.org/10.1007/978-3-030-33843-5_7

compressed sensing [4]. With parallel imaging techniques, the known sensitivities of multiple receive coils contribute to spatial encoding, and ultimately allow for an image reconstruction from sparser sampling. Compressed sensing reconstruction is an extension of traditional iterative reconstruction methods which estimate images from under-sampled data by enforcing consistency with acquired data and applying regularization – a model of a priori information about the reconstruction. In compressed sensing specifically, the regularization term enforces sparsity in some transform domain. Effective regularization is a key element for solving the under-sampled image reconstruction problem, however traditional regularization terms are often an oversimplification of MR image structure and offer limited a priori information.

Recently, machine learning based approaches for sparsely sampled image reconstruction were introduced [5–9]; some of these methods use a convolutional neural network to learn the regularization term of an iterative reconstruction [5, 7]. They were designed to generalize the concept of compressed sensing and learn the entire reconstruction procedure for multi-channel MR data. One such method is the Variational Network, which has been demonstrated for successful reconstruction of 4-fold accelerated knee images [7, 10], and 3-fold accelerated abdominal images [11].

The first objective of this work is to extend the VN approach to applications beyond knee and abdominal imaging and evaluate the performance of a VN jointly trained with data of multiple anatomical regions. The simplicity of a single network for a wide range of applications would be a substantial benefit for clinical workflow. The second objective is to evaluate a newly developed version of the VN which consists of a higher model capacity regularizer.

2 Methods

2.1 Image Acquisition

All scans were performed on a clinical 3T system (Siemens Magnetom Skyra), with different receive coils ranging from 12 to 26 elements. Fifty fully-sampled anatomical images were obtained from 5 anatomical areas, these areas – ranked in order of perceived image SNR – were brain, knee, hip, ankle, and shoulder. The study was approved by our institutional review board. The sequence parameters were as follows:

Ankle – Sagittal Fat-Saturated Proton-Density (PD-FS): TR = 2800 ms, TE = 30 ms, turbo factor (TF) = 5, matrix size = 384 × 384, in-plane resolution 0.42 × 0.42 mm^2, slice thickness = 3.0 mm.

Brain – Axial T2: TR = 6000 ms, TE = 113 ms, TF = 18, matrix size 384 × 384, in-plane resolution = 0.57 × 0.57 mm^2, slice thickness = 5.0 mm.

Hip – Coronal PD: TR = 3000 ms, TE = 32 ms, TF = 5, matrix size = 320 × 320, in-plane resolution = 0.5 × 0.5 mm^2, slice thickness = 3.0 mm.

Knee – Coronal PD: TR = 2750 ms, TE = 32 ms, TF = 4, matrix size = 320 × 320, resolution = 0.44 × 0.44 mm^2, slice thickness = 3.0 mm.

Shoulder – Coronal Fat-Saturated T2: TR = 4540 ms, TE = 54 ms, TF = 12, matrix size = 320 × 320, in-plane resolution = 0.44 × 0.44 mm^2, slice thickness = 3.0 mm.

The fully sampled images were then retrospectively under-sampled; the under-sampling was applied such that the center 24 lines of raw k-space data, and every fourth line beyond this center region were retained. The remaining k-space lines were set to zero. The center 24 lines were used for the ESPIRiT [12] estimation of coil sensitivities.

2.2 Variational Network

Experiments were performed with two versions of the VN. The first is the classic VN, described in Hammernik et al. [7] and the second is a version in which the regularizer is replaced with a Unet network [13] (Unet-VN).

For this study, we implemented the classic VN in Pytorch, and replaced the IPALM optimizer [14], which was traditionally used for VN training, with the Adam optimizer [15]. The regularizer in this network is a single convolutional layer with 48 11 × 11 convolutional kernels. The activation functions are a learned set of Gaussian radial basis functions, and the model capacity is approximately 131, 000 parameters.

In addition to the classic VN network, we also evaluated a Unet-VN network which was designed to have much higher model capacity (1.2 million parameters). For this architecture, we replace the regularizer in the classic model with a Unet network; otherwise the VN method was unchanged. Our Unet implementation has 3 encoding convolutional layers followed by 3 decoding convolutional layers, with 24, 48, 96, 48, 24, and 12 3 × 3 convolutional kernels respectively. Max-pooling and bi-linear interpolation were used for dimensionality reduction and expansion respectively. We used ReLU for the non-linear activation function, and instance normalization was applied during training.

2.3 Network Training

Individual trainings of the VN and Unet-VN were performed with 30 volumes of each anatomical region. Ten volumes for each dataset were reserved for a validation set and another 10 volumes were reserved for testing. Joint multi-anatomy training was performed with 6 volumes of each of the 5 anatomical regions for a total of 30 training cases. The Adam optimizer was used with a batch size of 1 and a learning rate of 3 × 10^{-4}. We used Mean squared error as the loss function. Convergence (validation loss stops decreasing) for each training was achieved at a different number of epochs ranging from 60 to 100. Training was performed on a Tesla P100 GPU.

2.4   Evaluation of Reconstructed Images

We tested the trained networks on data from 10 image volumes per anatomical region. These cases were not included in the training set. We compare the VN and Unet-VN reconstructions with the fully-sampled reference, the zero-filled reconstruction and a combined Parallel Imaging, Compressed Sensing reconstruction method based on Total Generalized Variation (PI-CS TGV) [16] For all of the PI-CS TGV reconstructions, the regularization parameter was set to 4×10^{-6} and the number of iterations was 1000. We compared the reconstruction results quantitatively in terms of structural similarity index (SSIM) [17].

Table 1. Structural similarity index was calculated for the 10 volumes in each test set; the mean and standard deviations are reported. Each of the 5 test sets were evaluated on all 12 trained networks. The row labels are the training sets used, and the column labels are the test set data.

	Mean structural similarity of predicted images				
	Brain	Knee	Ankle	Hip	Shoulder
VN training set					
Brain	0.976 (0.013)	0.965 (0.022)	0.948 (0.012)	0.948 (0.021)	0.830 (0.060)
Knee	0.971 (0.013)	0.974 (0.024)	0.947 (0.011)	0.948 (0.021)	0.843 (0.049)
Ankle	0.951 (0.010)	0.952 (0.012)	0.966 (0.006)	0.950 (0.017)	0.917 (0.023)
Hip	0.950 (0.009)	0.932 (0.013)	0.961 (0.007)	0.961 (0.013)	0.907 (0.027)
Shoulder	0.958 (0.010)	0.954 (0.016)	0.962 (0.006)	0.952 (0.017)	0.924 (0.020)
All	0.970 (0.012)	0.964 (0.019)	0.966 (0.006)	0.958 (0.015)	0.922 (0.020)
U net-VN training set					
Brain	0.979 (0.013)	0.942 (0.012)	0.957 (0.007)	0.933 (0.019)	0.876 (0.040)
Knee	0.968 (0.015)	0.981 (0.021)	0.956 (0.007)	0.945 (0.020)	0.890 (0.031)
Ankle	0.951 (0.013)	0.866 (0.021)	0.970 (0.005)	0.941 (0.018)	0.917 (0.024)
Hip	0.899 (0.025)	0.893 (0.023)	0.899 (0.025)	0.965 (0.012)	0.888 (0.026)
Shoulder	0.925 (0.023)	0.909 (0.011)	0.950 (0.013)	0.939 (0.020)	0.929 (0.019)
All	0.976 (0.014)	0.969 (0.017)	0.967 (0.005)	0.960 (0.015)	0.926 (0.019)

3   Results

The SSIM results for the VN and Unet-VN reconstructed images are reported in Table 1. We report the SSIM for all combinations of training and test data. For all anatomical regions, the highest SSIM is achieved with the individual, anatomy-specific, trained network. In these cases where the training and test anatomy are matched, the Unet-VN outperforms the classic VN. Image reconstruction results for the matched training and test sets are shown in Fig. 1. The VN and Unet-VN both outperform the PI-CS TGV method.

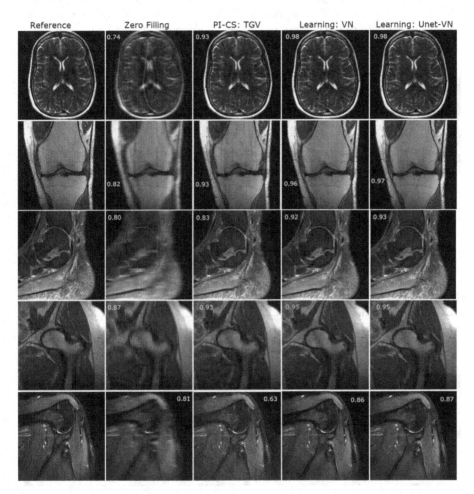

Fig. 1. Brain, knee, ankle, hip and shoulder reconstructions with 4-fold acceleration. The learned reconstructions appear sharper and have less residual artefacts than the PI-CS TGV reconstructions. The displayed SSIM values were calculated for the presented slices.

When the training data and test data are not matched we observe an increase in residual artefacts in the reconstructed image and a decrease in SSIM. This is demonstrated in Fig. 2 where we show knee images reconstructed with the VN individually trained with knee, brain, ankle and hip images. Another general trend that we observe when the training and test data is not matched is over-smoothing in the reconstructed images when training SNR < test SNR. When the opposite is true – training SNR > test SNR, noise amplification is observed. This effect is demonstrated in Fig. 3. When the training and test data are not matched, the classic VN outperforms the Unet-VN for the majority of the training set/test set combinations (16/20).

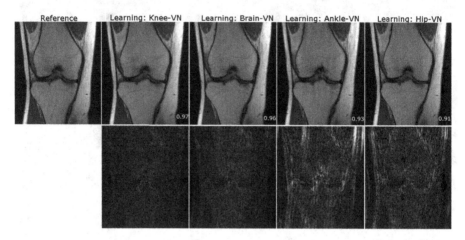

Fig. 2. Coronal PD weighted knee scan with 4-fold acceleration. The top row depicts the reconstructed results for the classic VN trained with knee, brain, ankle and hip images. The bottom row shows the difference images compared to the fully sampled reference.

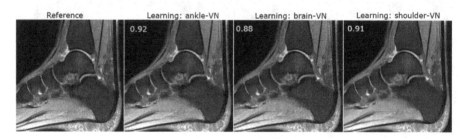

Fig. 3. Sagittal PD-FS ankle scan with 4-fold acceleration. Reconstruction results for the VN trained with ankle, brain (high SNR), and shoulder (low SNR). These results illustrate the trend that when training SNR > test SNR, the images suffer from noise amplification, and when training SNR < test SNR, the images appear over-smoothed.

The performance of the joint multi-anatomy trained networks approached that of the individual trainings for each anatomy and the Unet -VN consistently outperformed the classic VN. Image results for the multi-anatomy training are shown in Fig. 4.

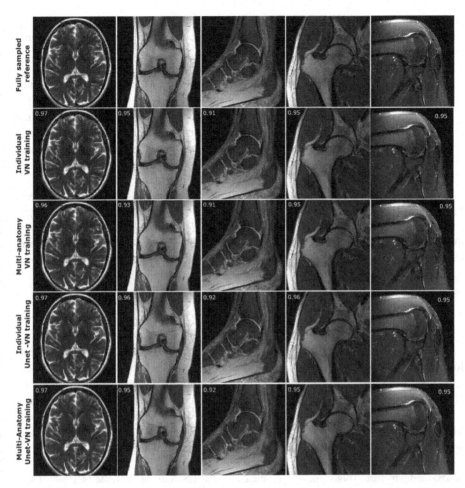

Fig. 4. Brain, knee, ankle, hip and shoulder reconstructions with 4-fold acceleration. The joint multi-anatomy trained networks result in similar reconstructed image quality as the individually trained networks. The Unet-VN matches or exceeds the classic VN for individual anatomy and multi-anatomy training.

4 Discussion

The variational network outperformed the PI-CS TGV algorithm for reconstructions of 4-fold accelerated knee, brain, hip, ankle and shoulder images. The Unet-VN which has a higher model capacity regularizer than the classic VN, outperforms the classic VN for individual trainings when the test and training data are matched. In addition to higher model capacity, the Unet regularizer – with multiple convolutional layers – has a larger receptive field than the classic single-layer regularizer. This may also contribute to the improved performance. The training time of the Unet-VN is approximately 25% longer than the training time of the classic VN. The Unet-VN network does not perform as well in most cases when the image being reconstructed is not represented in the training set, suggesting that the Unet-VN does not generalize as well to anatomical regions not previously seen by the network.

A specific trend is observed when there is a mismatch in the SNR of the training set and the test set; when the SNR of the training data is lower than the SNR of the test data, we observe over-smoothing in the reconstructed images. When the SNR of the training data is higher than the SNR of the test data, we see noise amplification in the reconstructed images. These findings are in agreement with a previous study that made a similar observation with fat-saturated (lower snr) and non – fat saturated (higher snr) knee images [10].

The networks that were jointly trained with multi-anatomy data have similar performance to those trained with a single anatomy, and again the Unet-VN outperforms the classic VN. A single network that can be used for many different clinical applications is not only beneficial for clinical workflow but also presents the opportunity for much larger training sets. In this study we used 30 images for joint multi-anatomy training in order to make fair comparisons with individual trainings; this approach does not take advantage of the 5x more training data that were available.

5   Conclusion

In this work, the classic VN and a newly developed Unet-VN were demonstrated for 4-fold acceleration of ankle, brain, hip and shoulder images and out-performed the PI-CS approach. The Unet-VN, with a higher model capacity regularizer, outperformed the classic VN for individual trainings as well as for joint multi-anatomy trainings. The networks jointly trained with multi-anatomy data had similar performance to those trained for a specific anatomy. Our findings suggest that the VN approach is a promising clinical tool for accelerated MR image reconstruction.

Acknowledgements. This work was supported by the US National Institute of Health under grants NIH/NIBIB R01EB024532 and P41EB017183 and the European Research Council starting grant HOMOVIS, No. 640156.

References

1. Griswold, M.A., et al.: Generalized autocalibrating partially parallel acquisitions (GRAPPA). Magn. Reson. Med. **47**(6), 1202–1210 (2002)
2. Pruessmann, K.P., et al.: SENSE: sensitivity encoding for fast MRI. Magn. Reson. Med. **42** (5), 952–962 (1999)
3. Sodickson, D.K., Manning, W.J.: Simultaneous acquisition of spatial harmonics (SMASH): fast imaging with radiofrequency coil arrays. Magn. Reson. Med. **38**(4), 591–603 (1997)
4. Lustig, M., Donoho, D., Pauly, J.M.: Sparse MRI: the application of compressed sensing for rapid MR imaging. Magn. Reson. Med. **58**(6), 1182–1195 (2007)
5. Aggarwal, H.K., Mani, M.P., Jacob, M.: MoDL: model-based deep learning architecture for inverse problems. IEEE Trans. Med. Imaging **38**(2), 394–405 (2019)

6. Akcakaya, M., et al.: Scan-specific robust artificial-neural-networks for k-space interpolation (RAKI) reconstruction: database-free deep learning for fast imaging. Magn. Reson. Med. **81** (1), 439–453 (2019)
7. Hammernik, K., et al.: Learning a variational network for reconstruction of accelerated MRI data. Magn. Reson. Med. **79**(6), 3055–3071 (2018)
8. Yang, G., et al.: DAGAN: deep de-aliasing generative adversarial networks for fast compressed sensing MRI reconstruction. IEEE Trans. Med. Imaging **37**(6), 1310–1321 (2018)
9. Zhu, B., et al.: Image reconstruction by domain-transform manifold learning. Nature **555** (7697), 487–492 (2018)
10. Knoll, F., et al.: Assessment of the generalization of learned image reconstruction and the potential for transfer learning. Magn. Reson. Med. **81**(1), 116–128 (2019)
11. Chen, F., et al.: Variable-density single-shot fast spin-echo MRI with deep learning reconstruction by using variational networks. https://doi.org/10.1148/radiol.2018180445
12. Uecker, M., et al.: ESPIRiT—an eigenvalue approach to autocalibrating parallel MRI: where SENSE meets GRAPPA. Magn. Reson. Med. **71**(3), 990–1001 (2014)
13. Ronneberger, O., Fischer, P., Brox, T.: U-Net: convolutional networks for biomedical image segmentation. In: Navab, N., Hornegger, J., Wells, William M., Frangi, Alejandro F. (eds.) MICCAI 2015. LNCS, vol. 9351, pp. 234–241. Springer, Cham (2015). https://doi.org/10. 1007/978-3-319-24574-4_28
14. Pock, T., Sabach, S.: Inertial proximal alternating linearized minimization (iPALM) for nonconvex and nonsmooth problems. SIAM J. Imaging Sci. **9**, 1756–1787 (2016)
15. Kingma, D.P., Ba, J.: Adam: A Method for Stochastic Optimization. arXiv:1412.6980 (2014)
16. Knoll, F., et al.: Second order total generalized variation (TGV) for MRI. Magn. Reson. Med. **65**(2), 480–491 (2011)
17. Wang, Z., et al.: Image quality assessment: from error visibility to structural similarity. IEEE Trans. Image Process. **13**(4), 600–612 (2004)

Modeling and Analysis Brain Development via Discriminative Dictionary Learning

Mingli Zhang[1][(✉)], Yuhong Guo[2], Caiming Zhang[3,4], Jean-Baptiste Poline[1], and Alan Evans[1]

[1] Montreal Neurological Institute, McGill University, Montreal, Canada
mingli.zhang@mcgill.ca
[2] School of Computer Science, Carleton University, Ottawa, Canada
[3] Shandong Co-innovation Center of Future Intelligent Computing, Yantai, China
[4] Shandong University of Finance and Economics, Jinan, China

Abstract. Research on modeling and exploring of the normal brain maturity, such as in vivo study of the anatomy of the developing brain, can provide references for developmental pathologies. In this paper, we model and explore brain development by learning a discriminative representation of the cortical brain data (T1 MRI) with a class-wise nonnegative dictionary learning (NDDL) approach. For each class, the proposed approach performs data modeling by first projecting the data into non-negative low-rank encoding coefficients with an analysis dictionary and then applying the coefficients onto an orthogonal synthesis dictionary to reconstruct the data. It also uses additional regularizers to enforce distal classes to fit into different analysis dictionaries. The learning problem is formulated as a sparse and low rank optimization problem, and solved with an alternating direction method of multipliers(ADMM). The effectiveness of the proposed approach is tested on brain age prediction problems by exploring the cortical status, and the experiments are conducted on the PING dataset. The proposed approach produces competitive results. Further, we were able for the first time to capture the status of brain thickness of specific cortical surface area with aging.

1 Introduction

Human brain development is a dynamic and complex process lasting through childhood, adolescence and adulthood. Modeling and exploring brain development is critical for diagnosis of neuropsychiatric disorder. Investigations on brain maturity (or brain age) have benefited from the development of advanced magnetic resonance imaging (MRI) [2] and from large-scale initiatives such as the Pediatric Imaging, Neurocognition, and Genetics (PING) [9] studies. Cross-sectional and longitudinal neuroimaging studies based on MRI have shown developmental trajectories of gray matter volumes, surface area and cortical thickness, contributing to a better understanding of brain maturation.

J.-B. Poline and A. Evans—Co-last author.

© Springer Nature Switzerland AG 2019
F. Knoll et al. (Eds.): MLMIR 2019, LNCS 11905, pp. 80–88, 2019.
https://doi.org/10.1007/978-3-030-33843-5_8

One of the simplest way to model brain maturity is predicting participant age from magnetic resonance imaging data through machine learning and statistical analysis. Modeling is usually done in two steps: (1) a statistical model is trained on lifespan data; (2) estimates of age are computed based on the trained model. By comparing the age predicted by the model with a participant's chronological age, one could infer a measure of the risk that an individual has on developing a neurodevelopmental disorder. The approach in [11] was able to increase the prediction accuracy by combining information from multiple brain imaging modalities. Studies show that combining multi-modal data can benefit the brain maturity prediction.

Representation learning techniques such as discriminative dictionary learning (DDL) are powerful algorithms to derive high-level latent features from high-dimensional and multi-modal data [6]. Most techniques (e.g., PCA and autoencoders) for this task do not take into account the class information during learning. DDL exploits the low rank and sparsity of high dimensional multi-modal data and can reduce feature dimension while preserving the significant information.

In this study, we propose a novel discriminative subspace learning approach called class-wise non-negative discriminative dictionary learning (NDDL) for modeling brain development data, by fusion of multi-feature brain imaging data into a common feature space. Our method differs from unsupervised subspace learning approaches like autoencoders [5] and the method dictionary projective learning [3,11]. In addition to exploring the low-rankness and non-negative nature of the class-specific projective features, we also take into account the biological significance variations across the neighboring classes in this model, by forcing the projective features from the remote-class to be sparse in each class-wise modeling. The major contributions of this work are as follows:

- **Novel framework:** We proposed framework of class-wise Non-negative Discriminative Dictionary Learning (NDDL). In addition to using Frobenius norm to push the sub-dictionary projection of samples from other classes to a nearly null space, we apply a sparse inducing l_1 norm to enforce the projections of samples from distal classes to be more sparse. To boost the discrimination of analysis dictionary $\mathbf{D}$ and projective features $\mathbf{A}$ and automatically determine the optimal dictionary size, we explore a weighted nuclear norm on $\mathbf{A}$ with non-negative constraint.
- **Clinical applications:** We evaluate the proposed NDDL method using the Pediatric Imaging, Neurocognition, and Genetics (PING) study data [9] on the task of modeling and exploring brain development. The proposed model is much more accurate in prediction compared with the state-of-the-art DPL. Our experiments conduct analysis on features of gray matter, which are important for predicting brain age, as well as the influence of gender on the prediction. The impact of cortical surface area, volumes and thickness on brain aging is also investigated with the proposed model. Last, we use our framework to explore cortical brain with aging.

2   The Proposed Approach

In this section, we present a class-wise non-negative discriminative dictionary learning (NDDL) method, which enforces l_1 norm regularizer on the encoding coefficients from distal classes, and applies weighted nuclear norm regularization and non-negative constraints on the class-wise encoding coefficient matrix.

2.1   Discriminative Dictionary Learning

We treat the modeling of brain development as a discriminative dictionary learning problem over K classes and each class is an age group. Let $\mathbf{X} = [\mathbf{X}_1, \cdots, \mathbf{X}_k, \cdots, \mathbf{X}_K]$ denote the data samples from all classes and $\mathbf{X}_k \in \mathbb{R}^{S \times N_k}$ denotes the data samples from the k-th class. Following [11] and [3], for each class k, we introduce an analysis dictionary $\mathbf{P}_k \in \mathbb{R}^{M \times S}$ to project the data into a coefficient matrix and then a synthesis dictionary $\mathbf{D}_k \in \mathbb{R}^{S \times M}$ to reconstruct the data, such that for all classes we have $\mathbf{P} = [\mathbf{P}_1, \cdots, \mathbf{P}_k, \cdots, \mathbf{P}_K]$ and $\mathbf{D} = [\mathbf{D}_1, \cdots, \mathbf{D}_k, \cdots, \mathbf{D}_K]$. With these dictionaries, one can perform data modeling with dictionary pair learning [3]:

$$\underset{\mathbf{P},\mathbf{D}}{\arg\min} \sum_{k=1}^{K} \|\mathbf{X}_k - \mathbf{D}_k \mathbf{P}_k \mathbf{X}_k\|_F^2 + \lambda \|\mathbf{P}_k \overline{\mathbf{X}}_k\|_F^2, \tag{1}$$

where $\| \cdot \|_F$ is the Frobenius norm, $\overline{\mathbf{X}}_k$ is the complementary data matrix of $\mathbf{X}_k$ in $\mathbf{X}$, in the format of $\overline{\mathbf{X}}_k = [\mathbf{X}_1, \cdots, \mathbf{X}_{k-1}, \mathbf{X}_{k+1}, \cdots, \mathbf{X}_K]$, and $\lambda > 0$ controls the trade-off between the reconstruction accuracy and regularization terms. The regularization term $\|\mathbf{P}_k \overline{\mathbf{X}}_k\|_F^2$ is used to forcing $\mathbf{P}_k \mathbf{X}_{\overline{k}}$ towards small or zero values for any other class $\overline{k} \in \{\overline{k} : |k - \overline{k}| \neq 0\}$. In this model, $\mathbf{P}_k$ projects the samples $\mathbf{X}_k$ into an encoding coefficient matrix $\mathbf{A}_k = \mathbf{P}_k \mathbf{X}_k$, it can reconstruct $\mathbf{X}_k$ with the synthesis dictionary $\mathbf{D}_k$.

The purpose of the Frobenius norm is to force the samples of other classes not to fit into the dictionary modeling of the current class and hence ensure the model to be class-wise discriminative. However, with the brain development analysis data, the class k has biological significance meanings. Data from neighboring classes (close age groups) can be similar and may share some modeling components in the dictionary learning. It might not be a good idea to penalize data from all other classes equally for each class k. Hence from each class k, we propose to only consider regularizations over data from its remote classes. We define the remote classes for class k as all the other classes $r(k)$ that are at least T steps away, such that $r(k) = \{\overline{k} : |k - \overline{k}| > T\}$, where T is a user-defined constant. Moreover, in addition to using Frobenius norm, we propose to use l_1 norm regularization, $\sum_{\overline{k} \in r(k)} \|\mathbf{P}_k \mathbf{X}_{\overline{k}}\|_1$, to push the projective coefficients of samples from the remote classes towards zeros.

Another important issue for dictionary pair learning is to determine the sizes/dimensions of the dictionaries, which can be tedious. Hence we propose to apply a low-rank regularization over the coefficient matrix produced by the analysis dictionary $\mathbf{P}_k$ with a weighted nuclear norm [4,10], $\|\mathbf{A}_k\|_{w,*}$ with

$\mathbf{A}_k = \mathbf{P}_k\mathbf{X}_k$. With this low-rank regularization, we can automatically push down the effective dimension of the dictionary $\mathbf{P}_k$. Correspondingly, we only need to maintain a compact synthesis dictionary $\mathbf{D}$ by enforcing an orthogonality constraint $\mathbf{D}_k^\top \mathbf{D}_k = \mathbf{I}$. These regularizations can also boost the discrimination of $\mathbf{D}$ and $\mathbf{A}$ by making each pair of dictionaries to compactly fit into the class-wise data.

Integrating all these components together, we have the following class-wise non-negative discriminative dictionary learning (NDDL) problem:

$$\underset{\mathbf{P},\mathbf{D}}{\arg\min} \sum_{k=1}^{K} \|\mathbf{X}_k - \mathbf{D}_k\mathbf{P}_k\mathbf{X}_k\|_F^2 + \sum_{\overline{k}\in r(k)} \left(\lambda\|\mathbf{P}_k\mathbf{X}_{\overline{k}}\|_F^2 + \lambda_1\|\mathbf{P}_k\mathbf{X}_{\overline{k}}\|_1\right) + \lambda_2\|\mathbf{P}_k\mathbf{X}_k\|_{w,*}$$

$$\text{s.t.} \quad \mathbf{D}_k^\top\mathbf{D}_k = \mathbf{I}, \ \mathbf{P}_k\mathbf{X}_k \geq 0, \ k = 1, ..., K. \tag{2}$$

where the first term of the objective is the reconstruction error, the second and third terms are regularizations over projections of samples from remote classes, and the forth regularization term enforces the representation coefficient matrix to be low rank to ensure a compact modeling on each class.

This joint minimization problem however is difficult to solve with different types of regularization terms. To facilitate the development of a relative easy training algorithm, for each class k we propose to introduce explicit encoding coefficient matrices $\mathbf{A}_k$ and $\{\mathbf{A}_{\overline{k}}\}$ with the equality constraints $\mathbf{A}_k = \mathbf{P}_k\mathbf{X}_k$ and $\mathbf{A}_{\overline{k}} = \mathbf{P}_k\mathbf{X}_{\overline{k}}$ for $\overline{k} \in r(k)$, which transforms the learning problem above into:

$$\underset{\mathbf{P},\mathbf{D},\{\mathbf{A}_k,\{\mathbf{A}_{\overline{k}}:\overline{k}\in r(k)\}\}}{\arg\min} \sum_{k=1}^{K}\|\mathbf{X}_k - \mathbf{D}_k\mathbf{P}_k\mathbf{X}_k\|_F^2 + \sum_{\overline{k}\in r(k)}\left(\lambda\|\mathbf{P}_k\mathbf{X}_{\overline{k}}\|_F^2 + \lambda_1\|\mathbf{A}_{\overline{k}}\|_1\right)$$

$$+ \lambda_2\|\mathbf{A}_k\|_{w,*}$$

$$\text{s.t.} \quad \mathbf{D}_k^\top\mathbf{D}_k = \mathbf{I}, \ \mathbf{A}_k = \mathbf{P}_k\mathbf{X}_k, \ \mathbf{A}_{\overline{k}} = \mathbf{P}_k\mathbf{X}_{\overline{k}}, \ \overline{k} \in r(k)$$

$$\mathbf{P}_k\mathbf{X}_k \geq 0, \ k = 1, ..., K. \tag{3}$$

To learn dictionary sets $\mathbf{D}$ and $\mathbf{P}$, we present an efficient optimization approach in the next section.

2.2 Training Algorithm

We propose to solve the model in (3) using an alternating direction method of multipliers (ADMM) based algorithm [1]. The principle of ADMM is to decompose a hard optimization problem into easier-to-solve sub-problems. We incorporate the equality constraints into the objective with auxiliary variable matrices $\{\mathbf{Z}_{1,k}\}$ and $\{\mathbf{Z}_{2,k}\}$. This leads to solve the following augmented Lagrangian function:

$$\underset{\mathbf{P},\mathbf{D},\{\mathbf{A}_k,\{\mathbf{A}_{\overline{k}}:\overline{k}\in r(k)\}\}}{\arg\min} \sum_{k=1}^{K} \|\mathbf{X}_k - \mathbf{D}_k\mathbf{P}_k\mathbf{X}_k\|_F^2$$

$$+ \sum_{\overline{k}\in r(k)} \left(\lambda\|\mathbf{P}_k\mathbf{X}_{\overline{k}}\|_F^2 + \lambda_1\|\mathbf{A}_{\overline{k}}\|_1 + \mu_2\|\mathbf{A}_{\overline{k}} - \mathbf{P}_k\mathbf{X}_{\overline{k}} + \mathbf{Z}_{2,k}\|_F^2\right)$$

$$+ \lambda_2\|\mathbf{A}_k\|_{w,*} + \mu_1\|\mathbf{A}_k - \mathbf{P}_k\mathbf{X}_k + \mathbf{Z}_{1,k}\|_F^2$$

$$\text{s.t.} \quad \mathbf{D}_k^\top\mathbf{D}_k = \mathbf{I}, \ \mathbf{A}_k = \mathbf{P}_k\mathbf{X}_k, \ \mathbf{A}_{\overline{k}} = \mathbf{P}_k\mathbf{X}_{\overline{k}}, \ \overline{k} \in r(k)$$
$$\mathbf{P}_k\mathbf{X}_k \geq 0, \ k = 1, ..., K. \tag{4}$$

In each iteration of the ADMM based algorithm, it alternatively updates each variable given the other variables are fixed as follows.

Updating P: Fixed $\mathbf{D}, \{\mathbf{A}_k, \{\mathbf{A}_{\overline{k}} : \overline{k} \in r(k)\}, \mathbf{Z}$, the minimization over each $\mathbf{P}_k$ is a quadratic minimization problem:

$$\underset{\mathbf{P}_k}{\arg\min} \quad \|\mathbf{X}_k - \mathbf{D}_k\mathbf{P}_k\mathbf{X}_k\|_F^2 + \sum_{\overline{k}\in r(k)} \left(\lambda\|\mathbf{P}_k\mathbf{X}_{\overline{k}}\|_F^2 + \mu_2\|\mathbf{A}_{\overline{k}} - \mathbf{P}_k\mathbf{X}_{\overline{k}} + \mathbf{Z}_{2,k}\|_F^2\right)$$

$$+ \mu_1\|\mathbf{P}_k\mathbf{X}_k - (\mathbf{A}_k + \mathbf{Z}_{1,k})\|_F^2 \tag{5}$$

which has the following closed-form solution:

$$\mathbf{P}_k = \left(\mathbf{D}_k^\top\mathbf{X}_k\mathbf{X}_k^\top + \mu_1(\mathbf{A}_k + \mathbf{Z}_{1,k})\mathbf{X}_k^\top + \sum_{\overline{k}\in r(k)} \left(\mu_2(\mathbf{A}_{\overline{k}} + \mathbf{Z}_{2,k})\mathbf{X}_{\overline{k}}^\top\right)\right)$$

$$\left((1+\mu_1)\mathbf{X}_k\mathbf{X}_k^\top + \sum_{\overline{k}\in r(k)} \left((1+\mu_2)\mathbf{X}_{\overline{k}}\mathbf{X}_{\overline{k}}^\top + \lambda\mathbf{X}_{\overline{k}}\mathbf{X}_{\overline{k}}^\top\right) + \gamma\mathbf{I}\right)^{-1} \tag{6}$$

where $\gamma = 10e^{-4}$ is a small constant used to increase invertibility.

Updating D: Given fixed $\mathbf{P}, \{\mathbf{A}_k, \{\mathbf{A}_{\overline{k}} : \overline{k} \in r(k)\}, \mathbf{Z}$, the minimization over each $\mathbf{D}_k$ is an orthogonal constrained optimization problem:

$$\underset{\mathbf{D}_k}{\arg\min} \quad \|\mathbf{X}_k - \mathbf{D}_k\mathbf{P}_k\mathbf{X}_k\|_F^2, \ \text{s.t.} \ \mathbf{D}_k^\top\mathbf{D}_k = \mathbf{I}$$

which can be rewritten as

$$\underset{\mathbf{D}_k}{\arg\min} \quad \text{tr}(\mathbf{D}_k^\top\mathbf{X}_k\mathbf{X}_k^\top\mathbf{P}_k^\top), \ \text{s.t.} \ \mathbf{D}_k^\top\mathbf{D}_k = \mathbf{I}, \tag{7}$$

Let $\mathbf{U}\mathbf{\Sigma}\mathbf{V}^\top$ be the singular value decomposition (SVD) of $\mathbf{X}_k\mathbf{X}_k^\top\mathbf{P}_k^\top$. (7) can be solved with $\mathbf{D}_k = \mathbf{U}\mathbf{V}^\top$.

Updating A: Given fixed $\{\mathbf{D}, \{\mathbf{A}_{\overline{k}} : \overline{k} \in r(k)\}, \mathbf{P}, \mathbf{Z}\}$, $\mathbf{A}_k$ is solved as follows:

$$\underset{\mathbf{A}_k}{\arg\min} \quad \|\mathbf{A}_k - (\mathbf{P}_k\mathbf{X}_k - \mathbf{Z}_k)\|_F^2 + \lambda_2\|\mathbf{A}_k\|_{w,*}, \text{s.t.} \ \mathbf{A}_k \geq 0, \tag{8}$$

$\mathbf{A}_k$ can be solved with weighted nuclear norm (WNN) [4,10] with the following constraint: $\mathbf{A}_k = \max(\mathbf{A}_k, 0)$.

Updating $\{\{\mathbf{A}_{\overline{k}} : \overline{k} \in r(k)\}$: Given fixed $\{\mathbf{D}, \mathbf{A}, \mathbf{P}, \mathbf{Z}\}$, we have a l_1 norm minimization problem over each $\overline{\mathbf{A}}_k$:

$$\underset{\{\mathbf{A}_{\overline{k}}:\overline{k}\in r(k)\}}{\arg\min} \sum_{\overline{k}\in r(k)} \left(\|\mathbf{A}_{\overline{k}} - (\mathbf{P}_k\mathbf{X}_{\overline{k}} - \mathbf{Z}_{2,\overline{k}})\|_F^2 + \lambda_1\|\mathbf{A}_{\overline{k}}\|_1 \right), \tag{9}$$

which has the following solution with element-wise soft-thresholding:

$$\{\{\mathbf{A}_{\overline{k}}:\overline{k}\in r(k)\}\}_i = \sum_{\overline{k}\in r(k)} \left(\mathbf{sign}([\mathbf{P}_k\mathbf{X}_{\overline{k}} - \mathbf{Z}_{2,k}]_i) \cdot \left([\mathbf{P}_k\mathbf{X}_{\overline{k}} - \mathbf{Z}_{2,k}]_i - \tfrac{\lambda_1}{\mu_2}\right)_+ \right) \tag{10}$$

where, $(\cdot)_+$ is $(x)_+ = x, x \geq 0$ and $(x)_+ = 0$ for others to any x.

Updating $\{\mathbf{Z}_k\}$: Finally, we update the dual variables following the standard ADMM algorithm: $\mathbf{Z}_{1,k} := \mathbf{Z}_{1,k} + (\mathbf{A}_k - \mathbf{P}_k\mathbf{X}_k)$ and $\mathbf{Z}_{2,k} := \mathbf{Z}_{2,k} + (\overline{\mathbf{A}}_k - \mathbf{P}_k\overline{\mathbf{X}}_k)$.

It is shown that, for sufficiently large values of ADMM parameters (i.e., μ_1 and μ_1) the algorithm is guaranteed to converge. In this paper, we set $T = 1$, $\mu_1 = 100$, $\mu_2 = 1$, $\lambda = 0.003$, $\lambda_1 = 0.001$ and $\lambda_2 = 0.0001$.

2.3   Classification

The learned dictionaries can be used to classify new samples by measuring the reconstruction error for each class. Considering the individual feature types (i.e., cortical surface area, thickness and volume). Let $\mathbf{x}^i \in \mathbb{R}^{S_i}$ be the features of type i for the sample to classify. We define as $e_k^i = \|\mathbf{x}^i - \mathbf{D}_k^i\mathbf{P}_k^i\mathbf{x}\|_2$ the error of reconstructing $\mathbf{x}^i$ with the dictionaries of class k for feature type i. We then assign the sample to the class whose dictionary gives the lowest error, i.e. $\widehat{k}_i = \arg\min_k e_k^i$.

To combine the information of multiple feature types, we use a subset of training examples (our validation set) to learn a regression model where inputs are the predicted ages $\widehat{k}_i$ for each feature type i and the output is the true subject age k_{real}.

$$\underset{\alpha}{\arg\min} \left(k_{\mathrm{real}} - \sum_i \alpha_i \widehat{k}_i \right)^2, \quad \text{s.t.} \sum_i \alpha_i = 1, \ \alpha_i \geq 0, \forall i. \tag{11}$$

Constraints on regression coefficients α_i enforce the final prediction to be a convex combination of predicted values for each feature type.

3   Experiments

The proposed NDDL framework is evaluated on modeling brain age and exploring brain development with structural MRI using measures of cortical thickness, surface area and volume and diffusion tensor imaging of 841 developing subjects. The 10-fold cross-validation is applied on this experiments. To measure performance in terms of prediction accuracy (ACC), root mean square error (RMSE) and mean absolute error (MAE) are applied in this paper. The experiments were conducted in Matlab R2017b using a i7-6700K CPU with 16 GB of RAM. We first evaluate our framework with the pre-processed PING data, MRI images were pre-processed using the CIVET[1] pipeline version 2.1.0. with the DKT protocol. DTI connectomes were derived with the NDMG pipeline, using the same acquisition protocol as [11].

3.1   Prediction of Brain Age

We first demonstrate the proposed approach's performance by predicting the brain age from children to adolescents ranging from 3 to 21 years old, based on cortical thickness, cortical surface area and cortical volumes of T1 structure MRI on the PING database. Here, we evaluate our method in a classification setting by dividing the 841 PING subjects (408 female, mean age: 12.52 years; 433 male, mean age: 12.58 years) into five age groups: preschool childhood (5–7 years), late childhood (8–10 years), early adolescence (11–13 years), middle adolescence (14–17 years) and late adolescence (18–21 years). We also evaluate the impact of subject sex by predicting the age of male and female subjects separately.

Table 1 compares the RMSE, MAE and accuracy obtained by our DDL approach to SVM and random forest (RF) [8] and that of recently proposed method DPL [3] for evaluating the impact of each constraint in the model. We see that our approach outperforms the DPL method [3]. $\mathbf{M.} - \|\mathbf{A}_k\|_{w,*}$ is our model without weighted nuclear norm constraint on $\mathbf{A}_k$ and similarly $\mathbf{M.} - \|\overline{\mathbf{A}}_k\|_1$ $\left(M. - \|\sum_{\overline{k} \in r(k)} \left(\mathbf{A}_{\overline{k}}\right)\|_1\right)$ is the proposed model without l_1 norm on $\mathbf{A}_{\overline{k}}$. From the Table 1, we can find the weighted nuclear norm constraint on $\mathbf{A}_k$ has a greater impact on results than the l_1 norm of $\mathbf{A}_{\overline{k}}$. The classification results on female data is much higher than male data. Compared to DPL [3], our approach yields improvements of about 0.055 in RMSE, 0.053 in MAE and 0.066 in accuracy.

Table 1. Classification results on the database of PING.

Method	RMSE			MAE			ACC		
	Male	Female	All	Male	Female	All	Male	Female	All
SVM	0.921	0.891	0.843	0.721	0.709	0.664	0.596	0.622	0.652
RF	0.916	0.873	0.860	0.745	0.704	0.690	0.679	0.686	0.716
DPL	0.826	0.854	0.757	0.666	0.675	0.612	0.666	0.727	0.710
M. - $\|\mathbf{A}_k\|_{w,*}$	0.762	0.753	0.747	0.609	0.603	0.575	0.755	0.757	0.769
M. - $\|\overline{\mathbf{A}}_k\|_1$	0.751	0.726	0.719	0.595	0.580	0.569	0.743	0.764	0.770
Ours	0.762	0.726	0.702	0.609	0.580	0.559	0.755	0.764	0.776

---

[1] http://www.bic.mni.mcgill.ca/ServicesSoftware/CIVET.

The classification performance with only one variable (i.e., cortical surface area, thickness, or cortical volume) is also evaluated in this section, the RMSE with surface area, thickness and volume are 1.160, 0.884 and 1.035, and the MAE are 0.823, 0.584 and 0.727, respectively. We see that the thickness is the most significant variable among the three variables, both in terms of RMSE and MAE.

3.2 Exploring the Cortical Brain Development

With the proposed model, we also can predict the development variables of the cortex with the trained $\mathbf{D}_k$ $\mathbf{P}_k$, $\mathbf{A}_k$ and one variable as input, k is the group class. Using the predicted variables (thickness), we explore the specific status of each surface area (parcellation). The *correlation coefficients* $(r(x, y))$, where x_i is vector of thickness in one parcel i sorted with increasing participant age, y is the vector of increasing participant age. Figure 1 shows the area significantly related to aging, the areas with smallest correlation coefficients $(r(x, y))$. We find that the parietal regions are the most significantly related to age, consistent with the report in [7].

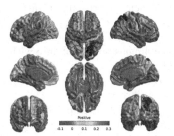 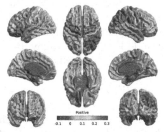

a) Tested significant thickness on specific   b) Predicted significant thickness on specific
surface area with increasing age.   surface area with increasing age.

Fig. 1. The significant negative r of thickness on sub-surface area with increasing age on 84 subjects

4   Conclusion

A non-negative discriminative dictionary learning model is proposed for predicting and modeling brain development. Compared with the conventional methods, this approach learns discriminative features by imposing both orthogonality on the synthesis dictionary, non-negativity low-rank constraints on projective coefficients, and a l_1 sparsity constraint on coefficients of non-class with biological boundary. An efficient alternative optimization algorithm based on ADMM was presented to learn the dictionaries with multivariate training data. Experiments on the tasks of predicting the brain age showed the benefit of our approach compared to state-of-the-art methods for these tasks. Furthermore, our approach can be used in understanding the influence of gender on brain development. With this model, we can capture the significant cortical surface area with aging.

Acknowledgements. This work was supported, in part, by the FRQS Quebec (CCC 246110, 271636), Brain Canada/HBHL/Exp (247858), National Nature Science Foundation of China (NSFC: 61902220, 61602277, U1609218), CONP RSA(201802), NIH-NIBIB P41 EB019936 (ReproNim) NIH-NIMH R01 MH083320 (CANDIShare) and NIH 5U24 DA039832 (NIF), as well as the Healthy Brains for Healthy Lives (HBHL) initiative.

References

1. Boyd, S., Parikh, N., Chu, E., Peleato, B., Eckstein, J.: Distributed optimization and statistical learning via the alternating direction method of multipliers. Found. Trends® Mach. Learn. **3**(1), 1–122 (2011)
2. Cole, J.H., Franke, K.: Predicting age using neuroimaging: innovative brain ageing biomarkers. Trends Neurosci. **40**, 681–690 (2017)
3. Gu, S., Zhang, L., Zuo, W., Feng, X.: Projective dictionary pair learning for pattern classification. In: Advances in Neural Information Processing Systems, pp. 793–801 (2014)
4. Gu, S., Zhang, L., Zuo, W., Feng, X.: Weighted nuclear norm minimization with application to image denoising. In: Proceedings of the IEEE Conference on Computer Vision and Pattern Recognition, pp. 2862–2869 (2014)
5. Hinton, G.E., Salakhutdinov, R.R.: Reducing the dimensionality of data with neural networks. Science **313**(5786), 504–507 (2006)
6. Li, J., Wu, Y., Zhao, J., Lu, K.: Low-rank discriminant embedding for multiview learning. IEEE Trans. Cybern. **47**(11), 3516–3529 (2017)
7. Sowell, E.R., Trauner, D.A., Gamst, A., Jernigan, T.L.: Development of cortical and subcortical brain structures in childhood and adolescence: a structural MRI study. Dev. Med. Child Neurol. **44**(1), 4–16 (2002)
8. Tong, T., et al.: A novel grading biomarker for the prediction of conversion from mild cognitive impairment to Alzheimer's disease. IEEE Trans. Biomed. Eng. **64**(1), 155–165 (2017)
9. Walhovd, K.B., et al.: Long-term influence of normal variation in neonatal characteristics on human brain development. Proc. Nat. Acad. Sci. **109**(49), 20089–20094 (2012)
10. Zhang, M., Desrosiers, C.: High-quality image restoration using low-rank patch regularization and global structure sparsity. IEEE Trans. Image Process. **28**(2), 868–879 (2019)
11. Zhang, M., Desrosiers, C., Guo, Y., Zhang, C., Khundrakpam, B., Evans, A.: Brain status prediction with non-negative projective dictionary learning. In: Shi, Y., Suk, H.-I., Liu, M. (eds.) MLMI 2018. LNCS, vol. 11046, pp. 152–160. Springer, Cham (2018). https://doi.org/10.1007/978-3-030-00919-9_18

Deep Learning for Computed Tomography

Virtual Thin Slice: 3D Conditional GAN-based Super-Resolution for CT Slice Interval

Akira Kudo[1]([✉]), Yoshiro Kitamura[1], Yuanzhong Li[1], Satoshi Iizuka[2], and Edgar Simo-Serra[3]

[1] Imaging Technology Center, Fujifilm Corporation, Minato, Tokyo, Japan
`akira.kudo@fujifilm.com`
[2] Center for Artificial Intelligence Research, University of Tsukuba, Tsukuba, Ibaraki, Japan
[3] Department of Computer Science and Engineering, Waseda University, Shinjuku, Tokyo, Japan

Abstract. Many CT slice images are stored with large slice intervals to reduce storage size in clinical practice. This leads to low resolution perpendicular to the slice images (*i.e.*, z-axis), which is insufficient for 3D visualization or image analysis. In this paper, we present a novel architecture based on conditional Generative Adversarial Networks (cGANs) with the goal of generating high resolution images of main body parts including head, chest, abdomen and legs. However, GANs are known to have a difficulty with generating a diversity of patterns due to a phenomena known as mode collapse. To overcome the lack of generated pattern variety, we propose to condition the discriminator on the different body parts. Furthermore, our generator networks are extended to be three dimensional fully convolutional neural networks, allowing for the generation of high resolution images from arbitrary fields of view. In our verification tests, we show that the proposed method obtains the best scores by PSNR/SSIM metrics and Visual Turing Test, allowing for accurate reproduction of the principle anatomy in high resolution. We expect that the proposed method contribute to effective utilization of the existing vast amounts of thick CT images stored in hospitals.

Keywords: Deep learning · Generative Adversarial Network · Super resolution · Computer vision · Computed tomography

1 Introduction

Image diagnosis plays an important role in recent healthcare solutions. The quality of diagnostic images largely affects the quality of diagnosis. The images such as CT or MRI acquired in hospitals are normally stored in Picture Archiving and Communication Systems (PACS). Although thin slice images, with slice intervals are about less than 1 mm, are frequently used for diagnosis, thick slice

© Springer Nature Switzerland AG 2019
F. Knoll et al. (Eds.): MLMIR 2019, LNCS 11905, pp. 91–100, 2019.
https://doi.org/10.1007/978-3-030-33843-5_9

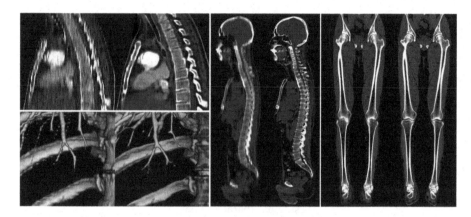

Fig. 1. Comparison of original thick image and the virtual thin image output generated by the proposed approach. On the left top row, the CT sagittal view of original thick image is blurred with each vertebrae bone being nearly indistinguishable, while they become clear in corresponding ×8 super resolution image (Virtual Thin). Arbitrary size data, even the whole body shown on the right, is available for inputs and capable of reconstructing natural image regardless of body part. On the left bottom row, fine blood vessel are smoothly reconstructed in a volume rendering view.

images with large slice intervals are used for long term storage to reduce the data size. However, the stored thick slice images do not have sufficient resolution for sagittal or coronal views, and also have limited applicability to 3D visualization (volume rendering). To address this, we present a novel super resolution algorithm for CT images based on Generative Adversarial Networks (GAN). Our goal is to generate high resolution 3D images corresponding from the input thick slice images. We base our approach on adversarial training [5] and aim to generate realistic-looking high-resolution CT images. One of the major difficulties is that CT images can be very diverse (*e.g.*, imaged body part, voxel size, resolution, slice thickness, slice interval, etc.), which can be difficult to synthesize with GANs. This difficulty is due to a phenomena known as mode collapse, in which the model becomes only able to synthesize a small subset of the original training data and presents a significant decrease in the output diversity [9]. We overcome this issue by additional conditioning of the discriminator on additional information, and use a three dimensional fully convolutional network to synthesize the high resolution CT images. Figure 1 shows example input thick images and the corresponding thin images synthesized by the proposed Virtual Thin Slice (VTS) method. The vertebrae bone structure is clearly reconstructed on the sagittal view, and fine blood vessels are reproduced well on the VR image.

2   Related Work

Single image super resolution is a major problems in computer vision field with a long history. The very first approach were filtering approaches, such as linear, bicubic or Lanczos filtering [4] which do not require huge computation. Yang et al. [14] categorized super resolution technique into 4 groups, which are prediction models, edge-based methods, image statistical methods and path-based methods. Recently, deep convolutional neural networks (CNN) based methods are showing significant performance in image recognition area [11]. SRCNN [3] improved performance on 2D image super resolution tasks by training non-linear low-resolution to high-resolution mappings using CNN filters. This is achieved through the minimization of pixel-wise Mean Squared Error (MSE) between reconstructed image and ground truth high resolution image. However, pixel-wise losses cannot capture perceptual differences [1], thus the output tends to be blurred and look unrealistic to human eyes. Adversarial training schemes [5] give much sharper result in image conversion task. Our work builds upon adversarial training to obtain better results. Some research focus on stabilize GAN training to reduce mode collapse, while there is a drawback of the computational cost [9].

Conditional GAN [10] is a conditioned min-max game between generator and discriminator. Isola [7] proposed Pix2Pix algorithm using images as conditional information, using pair of input image and target ground truth image. While adversarial approach gives more adequate to human perception, there are trade-offs between the perception and distortion of generating images [8].

In medical imaging, few researchers work on the 3D image super resolution. Chen et al. targeted 3D MRI super-resolution for medical image analysis [2]. Their target was limited to brain images. One of our contributions is realizing 3D CT image super resolution for any kind of body parts with a single generator network. Another contribution is the conditioning of the discriminator on the different body parts inspired by conditional GAN, and the ability to perform super-resolution of 3D medical images of arbitrary sizes.

3   Method

3.1   Objective Function

Our approach is based on conditional GAN, using pairs of low resolution data and high resolution data with slice information. The objective is to learn the transformation from of the thick slice image x to the virtual thin slice image y. Additionally, the discriminator is condition on a vector w, allowing the objective function to be expressed as

$$L_{cGAN}(G, D) = E_{(x,y)}\left[\log D(x, y, w)\right] + E_x\left[\log(1 - D(x, G(x), w))\right]. \quad (1)$$

where the model G tries to minimize this objective against an adversarial model D that tries to maximize it. Both G and D can be implemented as Convolutional Neural Networks (CNN). We also use a L_1 loss to calculate pixel-wise appearance differences between ground truth images and generated images, which has been shown to give less blurring than the L_2 loss in a diversity of image-to-image translation tasks [7]. Therefore, our final objective is expressed as

$$G^* = \arg\min_G \max_D L_{cGAN}(G, D) + \lambda L_{L1}(G) \quad (2)$$

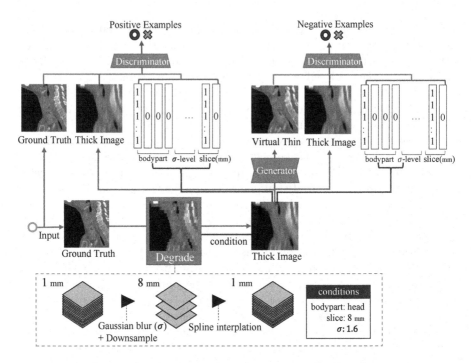

Fig. 2. Adversarial training framework of thick-thin slice translation on CT images. In each training iteration, the thin slice input data is randomly degraded to simulate thick slice data. For the Generator input, we feed 1 mm spline interpolated 3D thick slice image itself. On the other hand, for the Discriminator input, we feed generated Virtual Thin image with slice condition including body part information with degraded parameter scales.

Figure 2 illustrates the proposed adversarial training procedure. Note that the additional conditions are not inputted into the generator.

3.2   Network Architecture

Both our generator and discriminator models are based on Convolutional Neural Networks. Each convolutional layer consists of $4 \times 4 \times 4$ sized kernel followed by batch normalization [6] and LeakyReLU ($\alpha = 0.2$) as the activation function. Instead of max-pooling, strided convolution are used and image resolution is reduced to $1/2$ in each encoder convolutional layer. We set 64 channels for the first layer for both of the generator and the discriminator to get sufficient quality results and acceptable computation time. Figure 3 illustrates the architecture of our generator and discriminator networks.

Generator. The generator uses an encoder-decoder type architecture inspired by U-Net [12]. The resolution is decreased 4 times such that the minimum feature map size is $1/16$, and restored to the original size with trilinear interpolation. We used trilinear interpolation instead of transposed convolution for up-sampling to avoid generating checkered patterns due to uneven overlap of the kernel. In each convolutional layer, more feature channels generate better results, but require more resources and computational time. The generator estimates the high frequency components and the output is finally added to the input image.

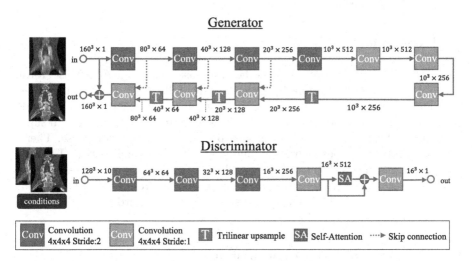

Fig. 3. Overview of the generator and discriminator network architectures. $(x^3 \times c)$ denotes the size of 3D feature map volumes, where c denotes the number of channels.

Discriminator. For the discriminator input, thick image (1 channel), thin image (1 channel) and slice information (8 channels) is given. Also a self-attention layer [15] is added in the fourth layer of the network. Self-attention mechanism is an idea to introduce global information between layers by computing attention maps which show the relevant area. In our case, self-attention did not largely affect the final performance, however, it speeded up the convergence of adversarial training. Final output is converted to a probability with a Sigmoid function.

3.3 Training Data

We introduce a Degrader procedure to randomly degrade the original thin image to thick slice image. In the Degrader, the input 3D image is down-sampled with Gaussian smoothing and spline interpolation is used to generate the missing slices. To simulate various combination of slice thickness and slice interval, the number of slices is reduced to either $1/4$ or $1/8$, then spline interpolation and random Gaussian noise is applied. The training samples are randomly cropped from the training images with an affine transform.

3.4   Conditioning Vector

We condition the discriminator on a vector w, containing various information about the input image. In particular, the type of input data (head, chest, abdomen or leg) is provided, in addition the slice interval (4 mm or 8 mm), and the scale of the standard deviation σ (2 scales) used for the Gaussian kernel which is treated as the slice thickness. In total, 8 channels are added as conditional information to the discriminator.

4   Experiments

For verification of the proposed method, we use Peak Signal-to-Noise Ratio (PSNR) and Structural Similarity Index Metric (SSIM) [13] as automatic metrics. There is a huge gap between PSNR and human perception sense, thus do not always present reasonable result. SSIM is more reasonable measure in super resolution field, however there is still a gap to human perception sense. Therefore, as additional experiment, we conducted Visual Turing Test (VTT). We asked 8 people who is either radiology technician or medical image research scientist to select the most high visibility image generated from 4 different methods each input. In the VTT, 4 images are shown in random order for 50 times to aggregate the answered ratio.

4.1   Datasets and Data Augmentation

We prepared 354 CT data (head:99, chest:98, abdomen:100, legs:57) for training which are obtained from diverse manufacturer's equipments (*e.g.*, GE, Siemens, Toshiba, etc.). They have been carefully selected to not contain metal artifacts or noises because the discriminator is prone to reproduce such artifacts. The input images CT values are clipped to the $[-2048, 2048]$ range and then normalized to be in the $[-1, 1]$ interval. In general, thick CT images are acquired in the range of 3–10 mm interval. On the other hand, 1 mm slice interval is enough for 3D visualization of principle anatomy by volume rendering. We set the experimental setting to generate 1 mm slice interval images from 8 mm. Therefore our datasets are only data with smaller than 1.0 mm slice interval. All images are rescaled to 1 mm isotropic voxels in preprocessing steps. In each training iteration, we randomly crop $160 \times 160 \times 160$ voxels from the input data and apply data augmentation. In particular, we apply an affine transformation consisting of a random rotation between $-5°$ and $+5°$, and random scaling between -5% and $+5\%$, both sampled from uniform distributions. The generated thick slice image inputs are subject to Gaussian filtering with σ uniformly sampled between 0.0 and 3.2 voxels before being downsampled to 1/4 or 1/8 resolution. As test datasets, we prepared 53 CT data (head:12, chest:16, abdomen:15, leg:10).

4.2   Results

We report for reference methods adapted to 3D, including bicubic, SRCNN [3], Pix2Pix [7], and Virtual Thin Slice (VTS, our approach). For our approach we

Table 1. PSNR and SSIM comparison result in experiments.

Methods	GANs	Conditional?	HF prediction?	PSNR	SSIM
Bicubic				32.34	0.878
SRCNN [3]				33.73	0.904
Pix2Pix [7]	✓			35.14	0.925
VTS (ours)	✓	✓	✓	**35.73**	**0.933**
(w/o) condition	✓		✓	35.17	0.924
(w/o) HF pred.	✓	✓		33.70	0.905
Ground Truth	–	–	–	∞	1.000

perform an ablative study where we remove the conditional vector from the discriminator and high-frequency component prediction. We employ SRCNN's 3 layer 9-1-5 model with each kernel expanded to 3D. Pix2Pix's convolutional networks are also replaced to $4 \times 4 \times 4$ sized kernels in each layer to adapt 3D and iterates down sampling until the feature image size become one pixel. Each type of network architecture is trained for around 100 epochs with Adam optimizer having learning rate for 2×10^{-4} and momentum parameter $\beta_1 = 0.5$.

Table 1 shows average PSNR and SSIM calculated over the test datasets. VTS has the highest score among other methods and the ablation study shows that removing either the conditional vector or high-frequency prediction lowers the quality of the generated outputs.

As we can see in Fig. 4, proposed VTS model generated the best perceptual quality with more sharpness and realistic images than other models. In particular, VTS works better with high intensity values such as bone boundary area rather than soft tissues. Although Pix2Pix model has similar PSNR/SSIM score to VTS, VTS was preferred roughly 90% of the time in VTT presented as shown in Fig. 5. The boxplot shows the answered ratio among 4 methods by the research participants. The images in which Pix2Pix was preferred over VTT consists primarily of legs data which have small difference between thick and thin as shown bottom row in Fig. 4. Even with some test data containing metal artifacts and unknown test patterns, the generated images are consistent with the input patterns, and don't contain enhancing artifacts or noise.

Another important feature of the proposed method is that the generator network is a fully convolutional neural network, and as such can handle each part of body and also field of view. Additionally, we have performed a verification test on 66 real thick slice images covering a wide condition of view with varying slice numbers (10 to 327), slice intervals (from 3.0 to 10.0 mm), and FOV (128 to 512 mm * mm). Using this dataset, we confirmed that the generator networks are able to successfully generate 1 mm slice interval images from the diversity of slice spacing and FOV images. We attribute these results to the wide range of data augmentation that we apply during training. Example of generated HR image from real existing thick slice image for either whole body or chest are shown in Fig. 1. The entire images are naturally reconstructed with no seams.

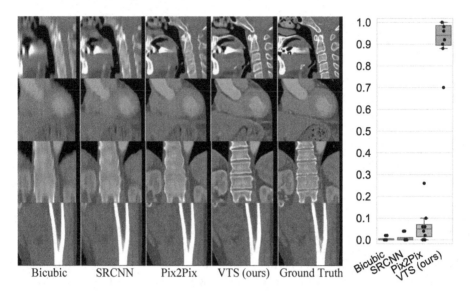

| Bicubic | SRCNN | Pix2Pix | VTS (ours) | Ground Truth |

Fig. 4. Comparison of the generated images of bicubic, SRCNN [3], pix2pix [7], VTS (ours) and corresponding ground truth thin slice image. [8× slice interpolation]

Fig. 5. The answered ratio in the Visual Turing Test.

5   Conclusion

In this paper, we have presented a super resolution algorithm that can be applicable for CT images of main body parts and various field of view. By inputting additional information regarding input data in the discriminator network, we show that output data quality increases significantly. Furthermore, the additional information is not necessary as test time. Numbering vertebrae bone is clearly easier with our VTS images compared to the original thick images. Also, we believe in-depth evaluation on abnormal images is an important next step for future work. In the future, we expect our VTS method will take on a role for the further development of medical image analysis and diagnosis support tasks, such as bone labeling and lung section segmentation, for thick slice data.

Acknowledgements. We acknowledge using the Reedbush-L (SGI Rackable C2112-4GP3/C1102-GP8) HPC system in the Information Technology Center, The University of Tokyo for the GPU computation required in this work.

Appendix

We include a variety of additional generated images from proposed VTS and other methods in Fig. 6.

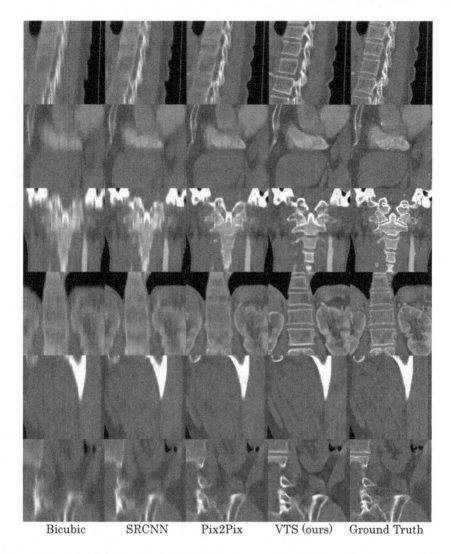

Bicubic   SRCNN   Pix2Pix   VTS (ours)   Ground Truth

Fig. 6. Results using bicubic, SRCNN [3], pix2pix [7], VTS (ours) and corresponding ground truth thin slice image.

References

1. Blau, Y., Michaeli, T.: The perception-distortion tradeoff. In: Proceedings of the IEEE Conference on Computer Vision and Pattern Recognition, pp. 6228–6237 (2018)
2. Chen, Y., Shi, F., Christodoulou, A.G., Xie, Y., Zhou, Z., Li, D.: Efficient and accurate MRI super-resolution using a generative adversarial network and 3D multi-level densely connected network. In: Frangi, A.F., Schnabel, J.A., Davatzikos, C., Alberola-López, C., Fichtinger, G. (eds.) MICCAI 2018. LNCS, vol. 11070, pp. 91–99. Springer, Cham (2018). https://doi.org/10.1007/978-3-030-00928-1_11

3. Dong, C., Loy, C.C., He, K., Tang, X.: Learning a deep convolutional network for image super-resolution. In: Fleet, D., Pajdla, T., Schiele, B., Tuytelaars, T. (eds.) ECCV 2014. LNCS, vol. 8692, pp. 184–199. Springer, Cham (2014). https://doi.org/10.1007/978-3-319-10593-2_13

4. Duchon, C.E.: Lanczos filtering in one and two dimensions. J. Appl. Meteorol. **18**(8), 1016–1022 (1979)

5. Goodfellow, I., et al.: Generative adversarial nets. In: Advances in Neural Information Processing Systems, pp. 2672–2680 (2014)

6. Ioffe, S., Szegedy, C.: Batch normalization: accelerating deep network training by reducing internal covariate shift. arXiv preprint arXiv:1502.03167 (2015)

7. Isola, P., Zhu, J.Y., Zhou, T., Efros, A.A.: Image-to-image translation with conditional adversarial networks. arXiv preprint (2017)

8. Ledig, C., et al.: Photo-realistic single image super-resolution using a generative adversarial network. arXiv preprint (2017)

9. Metz, L., Poole, B., Pfau, D., Sohl-Dickstein, J.: Unrolled generative adversarial networks. arXiv preprint arXiv:1611.02163 (2016)

10. Mirza, M., Osindero, S.: Conditional generative adversarial nets. arXiv preprint arXiv:1411.1784 (2014)

11. Radford, A., Metz, L., Chintala, S.: Unsupervised representation learning with deep convolutional generative adversarial networks. arXiv preprint arXiv:1511.06434 (2015)

12. Ronneberger, O., Fischer, P., Brox, T.: U-Net: convolutional networks for biomedical image segmentation. In: Navab, N., Hornegger, J., Wells, W.M., Frangi, A.F. (eds.) MICCAI 2015. LNCS, vol. 9351, pp. 234–241. Springer, Cham (2015). https://doi.org/10.1007/978-3-319-24574-4_28

13. Wang, Z., Bovik, A.C., Sheikh, H.R., Simoncelli, E.P.: Image quality assessment: from error visibility to structural similarity. IEEE Trans. Image Process. **13**(4), 600–612 (2004)

14. Yang, C.-Y., Ma, C., Yang, M.-H.: Single-image super-resolution: a benchmark. In: Fleet, D., Pajdla, T., Schiele, B., Tuytelaars, T. (eds.) ECCV 2014. LNCS, vol. 8692, pp. 372–386. Springer, Cham (2014). https://doi.org/10.1007/978-3-319-10593-2_25

15. Zhang, H., Goodfellow, I., Metaxas, D., Odena, A.: Self-attention generative adversarial networks. arXiv preprint arXiv:1805.08318 (2018)

Data Consistent Artifact Reduction for Limited Angle Tomography with Deep Learning Prior

Yixing Huang[1(✉)], Alexander Preuhs[1], Günter Lauritsch[2], Michael Manhart[2], Xiaolin Huang[3], and Andreas Maier[1,4]

[1] Pattern Recognition Lab, Friedrich-Alexander-Universität Erlangen-Nürnberg, 91058 Erlangen, Germany
yixing.yh.huang@fau.de
[2] Siemens Healthcare GmbH, 91301 Forchheim, Germany
[3] Institute of Image Processing and Pattern Recognition, Shanghai Jiao Tong University, Shanghai 200240, China
[4] Erlangen Graduate School in Advanced Optical Technologies (SAOT), 91058 Erlangen, Germany

Abstract. Robustness of deep learning methods for limited angle tomography is challenged by two major factors: (a) due to insufficient training data the network may not generalize well to unseen data; (b) deep learning methods are sensitive to noise. Thus, generating reconstructed images directly from a neural network appears inadequate. We propose to constrain the reconstructed images to be consistent with the measured projection data, while the unmeasured information is complemented by learning based methods. For this purpose, a data consistent artifact reduction (DCAR) method is introduced: First, a prior image is generated from an initial limited angle reconstruction via deep learning as a substitute for missing information. Afterwards, a conventional iterative reconstruction algorithm is applied, integrating the data consistency in the measured angular range and the prior information in the missing angular range. This ensures data integrity in the measured area, while inaccuracies incorporated by the deep learning prior lie only in areas where no information is acquired. The proposed DCAR method achieves significant image quality improvement: for 120° cone-beam limited angle tomography more than 10% RMSE reduction in noise-free case and more than 24% RMSE reduction in noisy case compared with a state-of-the-art U-Net based method.

Keywords: Deep learning · Limited angle tomography · Data consistency · Poisson noise · Robustness · Generalization ability

1 Introduction

Recently, deep learning has achieved overwhelming success in various computed tomography (CT) applications [1,2], including low-dose CT [3–5], sparse-view

© Springer Nature Switzerland AG 2019
F. Knoll et al. (Eds.): MLMIR 2019, LNCS 11905, pp. 101–112, 2019.
https://doi.org/10.1007/978-3-030-33843-5_10

reconstruction [6–8], and metal artifact reduction [9,10]. In this work, we are interested in the application of deep learning to limited angle tomography. Image reconstruction from data acquired in an insufficient angular range is called limited angle tomography. It arises when the gantry rotation of a CT system is restricted by other system parts, or a super short scan is preferred for the sake of quick scanning time, low dose, or less contrast agent.

Conventionally, limited angle tomography is addressed by extrapolation methods [11,12] or iterative reconstruction algorithms with total variation [13–15]. In the past three years, various deep learning methods have been investigated in limited angle tomography [16–21]. For example, Gu and Ye adapted the U-Net architecture [22] to learn artifacts from streaky images in the multi-scale wavelet domain [18]. Good quality images are obtained by this method for 120° limited angle tomography. The results presented in the literature reveal promising developments for a clinical applicability of deep learning-based reconstructions.

However, the robustness of deep learning in practical applications is still a concern. On one hand, deep learning methods may fail to generalize to new test instances as these methods are trained only on an insufficient dataset. On the other hand, due to the curse of high dimensional space [23], deep neural networks have been reported to be vulnerable to small perturbations, including adversarial examples and noise [24–26]. In the field of limited angle tomography, our previous work [19] has demonstrated that the U-Net method is not robust to Poisson noise as well. In this work, we devise an algorithm overcoming these limitations by enforcing data consistency with the measured raw data.

Since generating reconstructed images directly from a neural network appears inadequate, we propose to combine deep learning with known operators. The first category of such approaches is to build deep neural network architectures directly based on analytic formulas of conventional methods. In these neural networks, each layer represents a certain known operator whose weights are fine tuned by data-driven learning to improve precision. Therefore, they are called "precision learning" [27,28]. In precision learning, maximal error bounds are limited by prior information of the analytic formulas. Würfl et al. [16,20] proposed a neural network architecture based on filtered back-projection (FBP) to learn the compensation weights [29] for limited angle reconstruction. However, this particular method is not suitable for small angular ranges, e.g. 120° cone-beam limited angle tomography, since no redundant data are available to compensate missing data. The second category is to use deep learning and conventional methods to reconstruct different parts of an imaged object respectively. Bubba et al. [21] proposed a hybrid deep learning-shearlet framework for limited angle tomography, where an iterative shearlet transform algorithm [30] is utilized to reconstruct visible singularities of an imaged object while a U-Net based neural network with dense blocks [31] is utilized to predict the invisible ones. This method achieves better image quality than pure model or data-driven-based reconstruction methods. The third category is to use deep learning results as prior information for conventional methods. Zhang et al.'s method [10] is such an example for metal artifact reduction. To make the best of measured data, Zhang et al. used deep learning predictions as prior images to interpolate projection data in metal corrupted areas [10].

In this work, we choose the third category for limited angle tomography. In [19], the U-Net learns artifacts from streaky images in the image domain only. Reconstruction images obtained by such image-to-image prediction are very likely not consistent to measured data as the prediction does not have any direct connection to measured data. To make predicted images data consistent, a data consistent artifact reduction (DCAR) method is proposed: The predicted images are used as prior images to provide information in missing angular ranges first; Afterwards, a conventional reconstruction algorithm is applied to integrate the prior information in the missing angular ranges and constrain the reconstruction images to be consistent to the measured data in the acquired angular range.

2 Method

2.1 The U-Net Architecture

As displayed in Fig. 1, the same U-Net architecture as that in [19] is used for artifact reduction in limited angle tomography, which is modified from [22] and [18]. In this work, the input images are Ram-Lak-kernel-based FBP reconstructions from limited angle data, while the output images are artifact images. The Hounsfield scaled images are normalized to ensure stable training. An ℓ_2 loss function is used.

Fig. 1. The U-Net architecture for limited angle tomography (modified from [22]).

2.2 Data Consistent Artifact Reduction

Data Fidelity of Measured Data: We denote measured projections by p_m and the system matrix for the measured projections by A_m in cone-beam limited angle tomography. The FBP reconstruction from the measured data p_m only is denoted by f_{FBP}. The artifact image, predicted by the U-Net, is denoted by $f_{artifact}$. Then an estimation of the artifact-free image, denoted by f_{U-Net}, is obtained by $f_{U-Net} = f_{limited} - f_{artifact}$. Due to insufficient training data or sensitivity to noise in the application of limited angle tomography [19], f_{U-Net}

is not consistent to the measured data. A data consistent reconstruction image f follows the following constraint,

$$||A_m f - p_m|| < e_1, \tag{1}$$

where e_1 is a parameter for error tolerance. When the measured data p_m are noise-free, e_1 is ideally zero. When p_m contains noise caused by various physical effects, e_1 is a certain positive value.

Because of the severe ill-posedness of limited angle tomography, the number of images satisfying the above constraint is not unique. We aim to reconstruct an image which satisfies the above constraint and meanwhile is close the U-Net reconstruction $f_{\text{U-Net}}$. For this purpose, we choose to initialize the image f with $f_{\text{U-Net}}$ and solve it in an iterative manner, i.e.,

$$||A_m f - p_m|| < e_1, \text{ and } f^{(0)} = f_{\text{U-Net}}. \tag{2}$$

In this way, the data consistency constraint is fully satisfied. Note that with such initialization, the deep learning prior $f_{\text{U-Net}}$ contributes to the selection of one image among all images satisfying Eq. (1).

Data Fidelity of Unmeasured Data: We further denote the system matrix for an unmeasured angular range by A_u and its corresponding projections by p_u. In cone-beam computed tomography, a short scan is necessary for image reconstruction. Therefore, in this work, we choose A_u such that A_m and A_u form a system matrix for a short scan CT system. Although the projections p_u are not measured, they can be approximated by the deep learning reconstruction $f_{\text{U-Net}}$ via forward projection. Making the best of such prior information, the following constraint is proposed,

$$||A_u f - A_u f_{\text{U-Net}}|| = ||A_u(f - f_{\text{U-Net}})|| < e_2, \tag{3}$$

where the error tolerance parameter e_2 accounts for the inaccuracy of the deep learning prior $f_{\text{U-Net}}$. When $f_{\text{U-Net}}$ has bad image quality, a relative large value should be set. This constraint indicates that the final reconstruction f is close to the deep learning prior $f_{\text{U-Net}}$ in the unmeasured space and the difference between them is controlled by the parameter e_2.

Regularization: To further reduce noise and artifacts corresponding to the error tolerance of e_1 and e_2, additional regularization is applied. In this work, the following iterative reweighted total variation (wTV) regularization [15] is utilized,

$$||f^{(n)}||_{\text{wTV}} = \sum_{x,y,z} w^{(n)}_{x,y,z} ||\mathcal{D}f^{(n)}_{x,y,z}||,$$

$$w^{(n)}_{x,y,z} = \frac{1}{||\mathcal{D}f^{(n-1)}_{x,y,z}|| + \epsilon}, \tag{4}$$

where $f^{(n)}$ is the image at the nth iteration, $w^{(n)}$ is the weight vector for the nth iteration which is computed from the previous iteration, and ϵ is a small positive value added to avoid division by zero. A smaller value of ϵ results in finer image resolution but slower convergence speed.

Overall Algorithm: Therefore, the overall objective function for our DCAR method is as the following,

$$
\min \|\boldsymbol{f}\|_{\mathrm{wTV}}, \text{ subject to } \begin{cases} \boldsymbol{f}^{(0)} = \boldsymbol{f}_{\text{U-Net}}, \\ \|\boldsymbol{A}_{\mathrm{m}}\boldsymbol{f} - \boldsymbol{p}_{\mathrm{m}}\| < e_1, \\ \|\boldsymbol{A}_{\mathrm{u}}\boldsymbol{f} - \boldsymbol{A}_{\mathrm{u}}\boldsymbol{f}_{\text{U-Net}}\| < e_2, \end{cases} \tag{5}
$$

which is a constrained optimization problem.

To solve the above objective function, simultaneous algebraic reconstruction technique (SART) + wTV is applied [15], i.e., SART is utilized to minimize the data fidelity terms of Eqs. (1) and (3), while a gradient descent method is utilized to minimize the wTV term. To minimize the data fidelity terms, SART is adapted as the following,

$$
\boldsymbol{f}_j^{(l+1)} = \begin{cases} \boldsymbol{f}_j^{(l)} + \lambda \cdot \dfrac{\sum_{\boldsymbol{p}_i \in P_\beta} \frac{\mathcal{S}_{e_1}\left(p_i - \sum_{k=1}^{N} A_{i,k} \cdot f_k^{(l)}\right)}{\sum_{k=1}^{N} A_{i,k}} \cdot A_{i,j}}{\sum_{\boldsymbol{p}_i \in P_\beta} A_{i,j}}, & \text{if } \boldsymbol{p}_i \text{ is measured,} \\[3ex] \boldsymbol{f}_j^{(l)} + \lambda \cdot \dfrac{\sum_{\boldsymbol{p}_i \in P_\beta} \frac{\mathcal{S}_{e_2}\left(\sum_{k=1}^{N} A_{i,k} \cdot \left(f_{\text{U-Net}} - f_k^{(l)}\right)\right)}{\sum_{k=1}^{N} A_{i,k}} \cdot A_{i,j}}{\sum_{\boldsymbol{p}_i \in P_\beta} A_{i,j}}, & \text{otherwise,} \end{cases}
$$

$$\tag{6}$$

where the system matrix $\boldsymbol{A}$ is the combination of $\boldsymbol{A}_{\mathrm{m}}$ and $\boldsymbol{A}_{\mathrm{u}}$, the projection vector $\boldsymbol{p}$ is the combination of $\boldsymbol{p}_{\mathrm{m}}$ and $\boldsymbol{p}_{\mathrm{u}}$, and $\mathcal{S}_\tau$ is a soft-thresholding operator with threshold τ to deal with error tolerance. $\boldsymbol{p}_{\mathrm{u}}$ is estimated and substituted by $\boldsymbol{A}_{\mathrm{u}}\boldsymbol{f}_{\text{U-Net}}$ in the above formula. For other parameters, $\boldsymbol{f}_j$ stands for the jth pixel of $\boldsymbol{f}$, $\boldsymbol{p}_i$ stands for the ith projection ray of $\boldsymbol{p}$, $A_{i,j}$ is the element of $\boldsymbol{A}$ at the ith row and the jth column, l is the iteration number, N is the total pixel number of $\boldsymbol{f}$, λ is a relaxation parameter, β is the X-ray source rotation angle, and P_β stands for the set of projection rays when the source is at rotation angle β. To minimize the wTV term, the gradient of $\|\boldsymbol{f}\|_{\mathrm{wTV}}$ w.r.t. each pixel is computed and a gradient descent method using backtracking line search is applied [15].

2.3   Experimental Setup

We validate the proposed DCAR algorithm using 17 patients' data from the AAPM Low-Dose CT Grand Challenge [32] simulated in 120° cone-beam limited angle tomography without and with Poisson noise.

System Configuration: For each patient's data, limited angle projections are simulated in a cone-beam limited angle tomography system with parameters listed in Table 1. In the noisy case, Poisson noise is simulated considering an initial exposure of 10^5 photons at each detector pixel before attenuation.

Training and Test Data: To investigate the dependence of the U-Net's performance on training data, leave-one-out cross validation is performed. For each validation, data from 16 patients are used for training while the data from the

Table 1. The system configuration of cone-beam limited angle tomography to validate the proposed DCAR algorithm, where the angular parameters in the brackets are for a short scan configuration.

Parameter	Value
Scan angular range	$120°(210°)$
Start angle	$30°(0°)$
End Angle	$150°(210°)$
Angular step	$1°$
Source-to-detector distance	$1200.0\,\text{mm}$
Source-to-isocenter distance	$600.0\,\text{mm}$
Detector size	620×480
Detector pixel size	$1.0\,\text{mm} \times 1.0\,\text{mm}$
Image size	$256 \times 256 \times 256$
Image pixel size	$1.25\,\text{mm} \times 1.25\,\text{mm} \times 1.0\,\text{mm}$

remaining one are used for test. Among the 16 patients, 25 slices from each patient are chosen for training. For the validation patient, all the 256 slices from the FBP reconstruction f_{FBP} are fed to the U-Net for evaluation. As the artifacts are mainly caused by limited angle scan, the effect of cone-beam angle is neglected. Therefore, 2-D slices are used for training and test instead of volumes to avoid high computation. Both the training and test data are noise-free in the noise-free case, while both the training data and test data contain Poisson noise in the noisy case.

Algorithm Parameters: The U-Net is trained on the above data using the Adam optimizer. The learning rate is 10^{-3} for the first 100 epochs, 10^{-4} for the $101-130^{\text{th}}$ epochs, and 10^{-5} for the $131-150^{\text{th}}$ epochs. The ℓ_2-norm is applied to regularize the network weights. The regularization parameter is 10^{-4}.

For reconstruction, in the noise-free case, the error tolerance value e_1 is set to 0.001 in Eq. (1) for discretization error, while e_1 is set to 0.01 for the noisy case. The U-Net reconstructions $f_{\text{U-Net}}$ of each patient are reprojected in the angular range of $[0°, 210°]$. Other system parameters are the same as those in Table 1. A relatively large tolerance value of 0.5 is chosen empirically for e_2 in Eq. (3). For SART, the parameter λ in Eq. (6) is set to 0.8. For the wTV regularization, the parameter ϵ is set to 5 HU for weight update. 50 iterations of SART + wTV are applied using the U-Net reconstruction $f_{\text{U-Net}}$ as initialization to get the final reconstruction. For comparison, the results of 100 iterations of SART + wTV using zero images as initialization are presented.

3   Results

The results of three example slices in $120°$ noise-free cone-beam limited angle tomography are displayed in Fig. 2. These three slices are from Patient NO. 17,

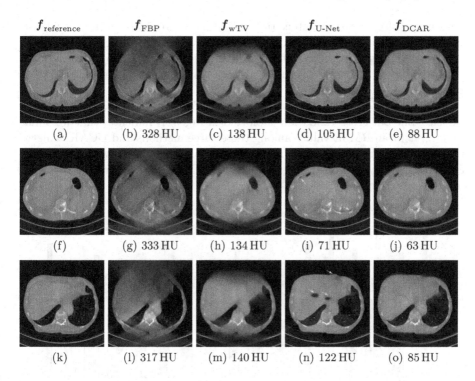

$f_{\text{reference}}$ f_{FBP} f_{wTV} $f_{\text{U-Net}}$ f_{DCAR}

(a)   (b) 328 HU   (c) 138 HU   (d) 105 HU   (e) 88 HU

(f)   (g) 333 HU   (h) 134 HU   (i) 71 HU   (j) 63 HU

(k)   (l) 317 HU   (m) 140 HU   (n) 122 HU   (o) 85 HU

Fig. 2. Reconstruction results of three example slices by U-Net and DCAR in noise-free 120° cone-beam limited angle tomography. The images from top to bottom are from Patient NO. 17, 4, and 7, respectively. The areas marked by the arrows are reconstructed incorrectly by the U-Net, which are rectified by DCAR. The RMSE value for each slice is displayed in their subtitle. Window: $[-1000, 1000]$ HU.

4, and 7, respectively. In each row, the reference image $f_{\text{reference}}$, the FBP reconstruction f_{FBP}, the U-Net reconstruction $f_{\text{U-Net}}$, the SART + wTV (using wTV for short in the following) reconstruction f_{wTV}, and the DCAR reconstruction f_{DCAR} are displayed in order. Comparing Fig. 4(b) with Fig. 4(a), the body outline of Patient 17 is severely distorted due to missing data. Moreover, many streaks occur, obscuring anatomical structures such as the ribs and the vertebra. Figures 4(c)–(e) demonstrate that wTV, U-Net, and DCAR all are able to improve these corrupted anatomical structures. The root-mean-square error (RMSE) is reduced significantly from 328 HU for f_{FBP} to 138 HU for f_{wTV} w.r.t. the reference image. But the intensity values at the top body part are still too low in f_{wTV}. The RMSE is further reduced to 105 HU for $f_{\text{U-Net}}$ in Fig. 4(d), while DCAR reaches the smallest RMSE value of 88 HU for this slice. In the middle row and the bottom row, the U-Net is able to reconstruct most anatomical structures well. However, the structures indicated by the red arrows are apparently incorrect compared with reference images. In Fig. 4(i), the dark holes indicated by the red arrows appear, very likely because the corresponding areas in Fig. 4(g) have low intensities due to dark streak artifacts. In contrast, the two large holes in Fig. 4(n) occur without any clear clue, since no dark areas are present in Fig. 4(l). Apparently Fig. 4(n) is not consistent to measured data and by using DCAR, these two holes are reestablished, although some darkness remains.

The comparison of the mean RMSE values for wTV, U-Net, and DCAR in the leave-one-out cross-validation is plotted in Fig. 3. It indicates that wTV has the largest mean RMSE values among these three methods and DCAR achieves more than 10% improvement in mean RMSE values compared with the U-Net. This convincingly demonstrates the benefit of DCAR in reducing artifacts for limited angle tomography.

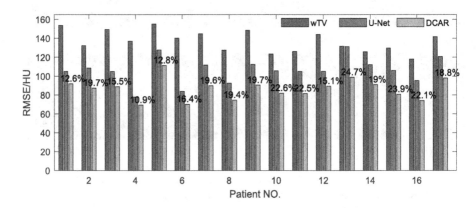

Fig. 3. Comparison of the mean RMSE values by wTV, U-Net, and DCAR for each patient in 120° noise-free cone-beam limited angle tomography. The relative improvement of DACR from the U-Net is marked for each patient.

In 120° cone-beam limited angle tomography with Poisson noise, the results of three example slices are displayed in Fig. 4. These three slices are from Patient NO. 17, 2, and 8, respectively. In the top row, Fig. 4(b) exhibits a high level of Poisson noise especially for the areas where a lot of X-rays are missing. The Poisson noise is entirely reduced by wTV in Fig. 4(c). However, like the noise-free cases, the top body area is still distorted. Figure 4(d) indicates that the U-Net trained on noisy data is still able to reduce limited angle artifacts. In addition, most Poisson noise is also prominently reduced and only a small portion of it remains. However, many low/median contrast structures, e.g. fat and muscles in the area marked by the red arrow, are blurred and cannot be distinguished between each other. Figure 4(e) indicates that DCAR can further reduce Poisson noise and improve low/median contrast structures, as no Poisson noise remains at all and the fat and muscle tissues can be distinguished between each other. The benefit of DCAR is also demonstrated by the RMSE value as it decreases from 138 HU in Fig. 4(d) to 102 HU in Fig. 4(e). For the slice in the middle row, the U-Net also reduces most of the artifacts and Poisson noise, comparing Fig. 4(i) with

$f_{\text{reference}}$ f_{FBP} f_{wTV} $f_{\text{U-Net}}$ f_{DCAR}

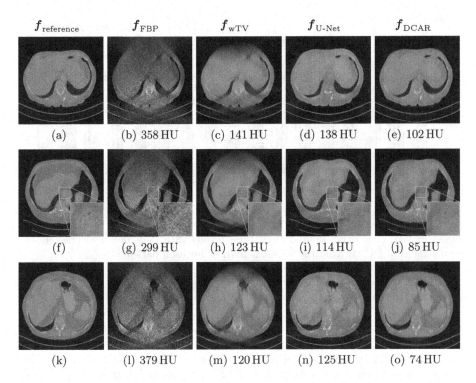

(a)   (b) 358 HU   (c) 141 HU   (d) 138 HU   (e) 102 HU

(f)   (g) 299 HU   (h) 123 HU   (i) 114 HU   (j) 85 HU

(k)   (l) 379 HU   (m) 120 HU   (n) 125 HU   (o) 74 HU

Fig. 4. Reconstruction results of three example slices by U-Net and DCAR in 120° cone-beam limited angle tomography with Poisson noise. The images from top to bottom are from Patient NO. 17, 2, and 8, respectively. The areas marked by the arrows are reconstructed incorrectly by the U-Net, which are rectified by DCAR. The RMSE value for each slice is displayed in their subtitle. Window: $[-1000, 1000]$ HU. (Color figure online)

Fig. 4(g). However, the cavities in the marked green box in Fig. 4(f) are missing in Fig. 4(i). They are smoothed out by the U-Net. Instead, DCAR is still able to reconstruct most of these cavities, as displayed in Fig. 4(j). For the slice in the bottom row, many dark dots occur in the U-Net reconstruction in Fig. 4(n), due to severe Poisson noise in the limited angle reconstruction in Fig. 4(l). However, these dark dots are eliminated by DCAR in Fig. 4(o). Except for these example slices, the comparison of the mean RMSE values for wTV, U-Net, and DCAR is displayed in Fig. 5. The mean RMSE values for wTV stay similar for both the noise-free and noisy cases. However, DCAR achieves more than 24% improvement compared with the U-Net in the noisy case. These remarkable results have demonstrated the robustness of DCAR to Poisson noise in 120° cone-beam limited angle tomography.

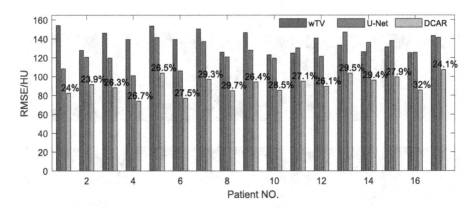

Fig. 5. Comparison of the mean RMSE values by wTV, U-Net, and DCAR for each patient in 120° cone-beam limited angle tomography with Poisson noise. The relative improvement of DCAR from the U-Net is marked for each patient.

4   Discussion and Conclusion

In the cross-validation experiments, for each test, 16 patients' CT data are used to train the U-Net. Since only 13 slices are chosen from each patient, 400 slices in total are used for training, which is very likely insufficient. Therefore, the U-Net training on such data has a limited generalization ability to test data. That is one potential cause to the dark holes in the U-Net reconstructions in Fig. 2 in the noise-free case. The occurrence of such dark holes make deep learning reconstructions not consistent to measured projection data. DCAR has the ability to improve such reconstructions by constraining them consistent to measured data.

In the noisy case, due to the curse of high dimensional space, noise will accumulate at each layer of the U-Net. Therefore, even if noise has a small magnitude, it still has a severe impact on the output images. That is why the U-Net is not robust to Poisson noise [19]. In this work, the U-Net is trained on data with Poisson noise. This endows the U-Net to deal with Poisson noise to a certain degree. Figure 4 indicates that the U-Net is able to reduce a certain level of Poisson noise in a manner of smoothing structures. In such a manner, some fine structures are also smoothed out, e.g., the small cavities in Fig. 4(f). In addition, in our experimental setup for the noisy case, the initial photon number without attenuation is relatively low. Hence, the Poisson noise in the FBP reconstruction images is well observed. In some cases, e.g. in Fig. 4(g), the Poisson noise is so strong that the U-Net is not able to reduce it. However, DCAR adapts the SART algorithm using soft-thresholding operators, which is noise tolerant. In addition, the wTV regularization further reduces the influence of Poisson noise as such high frequency noise pattern contradicts a gradient-sparse image, which wTV seeks.

In conclusion, the proposed DCAR method has better generalization ability to unseen data and is more robust to Poisson noise than the U-Net. This is demonstrated by our experiments, achieving significant image quality improvement. Compared to the U-Net, our method reduces the RMSE by more than 10% in the noise-free case and 24% in the noisy case for 120° cone-beam limited angle tomography.

Disclaimer. The concepts and information presented in this paper are based on research and are not commercially available.

References

1. Maier, A., Syben, C., Lasser, T., Riess, C.: A gentle introduction to deep learning in medical image processing. Zeitschrift für Medizinische Physik **29**(2), 86–101 (2019)
2. Wang, G., Ye, J.C., Mueller, K., Fessler, J.A.: Image reconstruction is a new frontier of machine learning. IEEE Trans. Image Process. **37**(6), 1289–1296 (2018)
3. Kang, E., Min, J., Ye, J.C.: A deep convolutional neural network using directional wavelets for low-dose X-ray CT reconstruction. Med. Phys. **44**(10), e360–e375 (2017)
4. Chen, H., et al.: Low-dose CT via convolutional neural network. Biomed. Opt. Express **8**(2), 679–694 (2017)
5. Yang, Q., et al.: Low dose CT image denoising using a generative adversarial network with Wasserstein distance and perceptual loss. IEEE Trans. Med. Imaging **37**, 1348–1357 (2018)
6. Han, Y.S., Yoo, J., Ye, J.C.: Deep residual learning for compressed sensing CT reconstruction via persistent homology analysis. arXiv preprint (2016)
7. Han, Y., Ye, J.C.: Framing U-Net via deep convolutional framelets: application to sparse-view CT. IEEE Trans. Med. Imaging **37**(6), 1418–1429 (2018)
8. Chen, H., et al.: LEARN: learned experts' assessment-based reconstruction network for sparse-data CT. IEEE Trans. Med. Imaging **37**(6), 1333–1347 (2018)
9. Gjesteby, L., Yang, Q., Xi, Y., Zhou, Y., Zhang, J., Wang, G.: Deep learning methods to guide CT image reconstruction and reduce metal artifacts. In: Medical Imaging 2017: Physics of Medical Imaging, vol. 10132, p. 101322W. International Society for Optics and Photonics (2017)
10. Zhang, Y., Yu, H.: Convolutional neural network based metal artifact reduction in X-ray computed tomography. IEEE Trans. Med. Imaging **37**(6), 1370–1381 (2018)
11. Louis, A.K., Törnig, W.: Picture reconstruction from projections in restricted range. Math. Methods Appl. Sci. **2**(2), 209–220 (1980)
12. Huang, Y., et al.: Restoration of missing data in limited angle tomography based on Helgason-Ludwig consistency conditions. Biomed. Phys. Eng. Express **3**(3), 035015 (2017)
13. Sidky, E.Y., Pan, X.: Image reconstruction in circular cone-beam computed tomography by constrained, total-variation minimization. Phys. Med. Biol. **53**(17), 4777 (2008)
14. Chen, Z., Jin, X., Li, L., Wang, G.: A limited-angle CT reconstruction method based on anisotropic TV minimization. Phys. Med. Biol. **58**(7), 2119 (2013)
15. Huang, Y., Taubmann, O., Huang, X., Haase, V., Lauritsch, G., Maier, A.: Scale-space anisotropic total variation for limited angle tomography. IEEE Trans. Radiat. Plasma Med. Sci. **2**(4), 307–314 (2018)

16. Würfl, T., Ghesu, F.C., Christlein, V., Maier, A.: Deep learning computed tomography. In: Ourselin, S., Joskowicz, L., Sabuncu, M.R., Unal, G., Wells, W. (eds.) MICCAI 2016. LNCS, vol. 9902, pp. 432–440. Springer, Cham (2016). https://doi.org/10.1007/978-3-319-46726-9_50

17. Zhang, H., et al.: Image prediction for limited-angle tomography via deep learning with convolutional neural network. arXiv preprint (2016)

18. Gu, J., Ye, J.C.: Multi-scale wavelet domain residual learning for limited-angle CT reconstruction. In: Proceedings of Fully 3D, pp. 443–447 (2017)

19. Huang, Y., Würfl, T., Breininger, K., Liu, L., Lauritsch, G., Maier, A.: Some investigations on robustness of deep learning in limited angle tomography. In: Frangi, A.F., Schnabel, J.A., Davatzikos, C., Alberola-López, C., Fichtinger, G. (eds.) MICCAI 2018. LNCS, vol. 11070, pp. 145–153. Springer, Cham (2018). https://doi.org/10.1007/978-3-030-00928-1_17

20. Würfl, T., et al.: Deep learning computed tomography: learning projection-domain weights from image domain in limited angle problems. IEEE Trans. Med. Imaging 37(6), 1454–1463 (2018)

21. Bubba, T.A., et al.: Learning the invisible: a hybrid deep learning-shearlet framework for limited angle computed tomography. Inverse Probl. 35(6), 064002 (2019)

22. Ronneberger, O., Fischer, P., Brox, T.: U-Net: convolutional networks for biomedical image segmentation. In: Navab, N., Hornegger, J., Wells, W.M., Frangi, A.F. (eds.) MICCAI 2015. LNCS, vol. 9351, pp. 234–241. Springer, Cham (2015). https://doi.org/10.1007/978-3-319-24574-4_28

23. Goodfellow, I.J., Shlens, J., Szegedy, C.: Explaining and harnessing adversarial examples. arXiv preprint (2014)

24. Szegedy, C., et al.: Intriguing properties of neural networks. arXiv preprint (2013)

25. Yuan, C., He, P., Zhu, Q., Bhat, R., Li, X.: Adversarial examples: attacks and defenses for deep learning. arXiv preprint (2017)

26. Antun, V., Renna, F., Poon, C., Adcock, B., Hansen, A.C.: On instabilities of deep learning in image reconstruction-does AI come at a cost? arXiv preprint arXiv:1902.05300 (2019)

27. Syben, C., et al.: Precision learning: reconstruction filter kernel discretization. In: Noo, F. (ed.) Procs CT Meeting, pp. 386–390 (2018)

28. Maier, A.K., et al.: Learning with known operators reduces maximum training error bounds. arXiv preprint (2019). Paper conditionally accepted in Nature Machine Intelligence

29. Riess, C., Berger, M., Wu, H., Manhart, M., Fahrig, R., Maier, A.: TV or not TV? That is the question. In: Proceedings Fully 3D, pp. 341–344 (2013)

30. Frikel, J.: Sparse regularization in limited angle tomography. Appl. Comput. Harmon. Anal. 34(1), 117–141 (2013)

31. Huang, G., Liu, Z., Van Der Maaten, L., Weinberger, K.Q.: Densely connected convolutional networks. Proc. CVPR 1(2), 3 (2017)

32. McCollough, C.H., et al.: Low-dose CT for the detection and classification of metastatic liver lesions: results of the 2016 low dose CT grand challenge. Med. Phys. 44(10), e339–e352 (2017)

Measuring CT Reconstruction Quality with Deep Convolutional Neural Networks

Mayank Patwari[1,2](✉)(iD), Ralf Gutjahr[2], Rainer Raupach[2], and Andreas Maier[1,3]

[1] Friedrich-Alexander Universität Erlangen-Nürnberg, 91058 Erlangen, Germany
[2] Siemens Healthcare GmbH, 91301 Forchheim, Germany
mayank.patwari@siemens-healthineers.com
[3] Erlangen Graduate School in Advanced Optical Technologies (SAOT), 91058 Erlangen, Germany

Abstract. With the increasing use of CT in diagnostic imaging, reducing the clinical radiation dose is necessary for ensuring patient safety. Reduced radiation dose results in quantum noise which adversely affects image quality and diagnostic value. Moreover, obtaining high quality images to act as reference images for image quality assessment is difficult. Therefore, automatic no-reference quality assessment of reconstructed images is necessary to preserve diagnostic image quality, while controlling radiation dose. In this work, we investigate the use of a deep convolutional neural network to measure CT image quality. Our developed metric shows concordance with conventional metrics of CT image quality ($|r| > 0.75$, $|\rho| > 0.75$). Our metric ranks images in terms of quality highly accurately ($\tau = 0.98$). We measure noise textures and levels not present in our training dataset. Furthermore, the proposed metric shows the improved quality in high dose iteratively reconstructed images, and the reduced quality in low dose images.

Keywords: Quantum noise · Convolutional neural network · Computed tomography · Image quality

1 Background

Computed tomography (CT) is one of the most important tools in medical imaging. CT is used in the diagnosis of both malignant and benign lesions in the brain, lungs, and liver, among other applications. Necessary information for tumor diagnosis can be affected by the reconstruction quality of the CT image. Because of the ALARA principle to improve patient safety, reconstructed CT images contain quantum noise. Quantum noise manifests itself as dark streaks in an image [10,11]. Large amounts of quantum noise may remove small and low-contrast tumors from a reconstructed image. With the increase in the use of CT for diagnostic purposes, automated measures of CT reconstruction quality and quantum noise content are becoming necessary to maintain high diagnostic value.

© Springer Nature Switzerland AG 2019
F. Knoll et al. (Eds.): MLMIR 2019, LNCS 11905, pp. 113–124, 2019.
https://doi.org/10.1007/978-3-030-33843-5_11

The noise power spectrum (NPS) has been used as a method to measure image noise content [3]. Successful calculation of the noise power spectrum requires an assumption of local signal linearity [17]. The NPS can be applied on homogenous phantoms such as water phantoms; however, assumptions of local linearity and homogeneity are violated in clinical CT images. The inhomogeneous nature of a clinical CT image and the non-stationary nature of quantum noise make the NPS an impractical choice for quantifying reconstruction quality [17].

Current metrics of CT reconstruction quality include the structured similarity index (SSIM) [18], peak signal to noise ratio (PSNR), and mean absolute error/mean squared error (MAE/MSE). The SSIM and its variants [1,6,19] have shown good agreement with radiologists on the quality of radiological images [13]. The PSNR and the MAE/MSE do not account for the overall quality of the image. The above metrics have all been used to quantify the effectiveness of CT denoising algorithms [7,22–24]. However, these metrics are full reference metrics and require the ground truth image to be present. In clinical CT, this requirement is typically difficult to achieve. Low noise reference CT images are typically produced using high radiation dose. Acquiring such a ground truth image for a clinical scan requires exposing the patient to a high dose of radiation, which endangers the patient's safety. Therefore, the development of a new metric which provides an automatic no-reference assessment of CT noise and reconstruction quality is desirable.

Recently, interest has grown in the application of deep learning for the development of no-reference image quality metrics. Kim et al. [4] used a convolutional neural network (CNN) to measure image distortion and quality without a reference image for digital images. Deep learning based quality metrics have also been applied to medical imaging problems by Galdran et al. [2], who proposed the use of a CNN to estimate the quality of retinal image segmentation. In this study, we use a CNN to learn a similarity score between a reference image and a reconstructed image with a significant amount of quantum noise. The learned similarity can be considered as a quality reference score for the noisy image.

2 Methodology

2.1 Data

The volumes were reconstructed from body CT scans of seven patients. The data was partitioned into training, holdout, and test sets. The training set contains 175 slices from Patient 1, 151 slices from Patient 2, 142 images from Patient 3, and 145 images from Patient 4. The holdout set contains 85 slices from Patient 1, 75 slices from Patient 2, 70 slices from Patient 3, and 71 slices from Patient 4. The test set contains 203 slices from Patient 5, 204 slices from Patient 6, and 453 slices from Patient 7. We use the holdout to observe the intra-patient generalization of our network. We use the test set to observe the inter-patient performance of our network. An abdominal window with a center value of 40

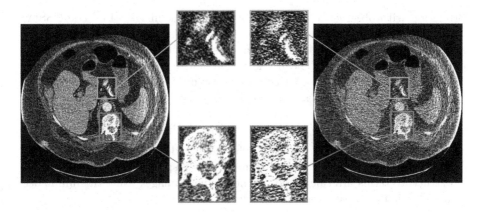

Fig. 1. A reconstructed full dose image (left) with a reconstructed half dose image (right) showing noise affected changes. This image was reconstructed with an abdominal window (center = 40 HU, width = 300 HU).

HU and a width of 300 HU was applied to all slices. The slices were 256×256 in size. The training dataset was augmented 8 times by rotations (2) and flips (4).

We generate noisy and denoised images from the same set of measured projections. We add quantum noise to a reconstruction by simulating a decrease in the applied dose with which the data was collected. The dose reduction is simulated by adding Poisson noise to the pre-log X-ray projections to simulate quantum noise. The Poisson noise is approximated by the following equation:

$$P(n) = \frac{\overline{n}^n}{n!} e^{-\overline{n}} \tag{1}$$

where n is the number of photons involved in transmission and $\overline{n}$ is the average number of photons detected at the detector. Lower radiation dose lowers the number of photons that are transmitted and therefore lowers the value of denominator, increasing the Poisson noise applied, and vice versa. The noise is added to the pre-log projections in post-processing, which makes it possible for us to generate multiple images with differing noise contents and differing noise structures from the same set of projections (See Fig. 1). No noise was added for the full dose images.

The volumes are reconstructed from the raw data using the weighted filtered backprojection (WFBP) algorithm [16]. Full dose images, which we use as reference images, are obtained by reconstructing with no dose reduction. Half dose and quarter dose images are obtained by simulating a dose reduction of 50% and 75% for the same raw dataset respectively.

2.2   Quality Estimation with Gradient Structural Similarity

We use the Gradient Structural Similarity Index (GSSIM) [1] as a proxy for image quality instead of the standard SSIM as the former shows greater correlation with observer assessments for radiological images [13]. The GSSIM is calculated for the reduced dose images by comparison with the corresponding full dose images. The GSSIM is based on the sensitivity of the human visual system to edge based artifacts. The gradient of an image I is calculated by applying the Sobel filters in the X and Y directions. The image $I_g = |G_x(I)| + |G_y(I)|$ is the gradient image. The GSSIM is calculated by the following formula:

$$GSSIM(I, I_r) = l(I, I_r)c_g(I, I_r)s_g(I, I_r) \qquad (2)$$

where I is the full dose image, and I_r is the reduced dose image. The l, c, s terms are the luminance, contrast and structure terms respectively [1]. The contrast and structure terms are calculated on the gradient images while the luminance term is calculated on the grayscale image.

2.3 Image Normalization

Since the human visual system focuses on high frequency errors, we filter out the low frequency components of the image. We implement the filtering scheme used by Kim et al. [4]. The high pass filtering is carried out by subtracting a low resolution version of the image I from the image. The low resolution image $\bar{I}$ is created by Gaussian blurring and downsampling by 2. The downscaled image is upsampled to the original image resolution to match image dimensions.

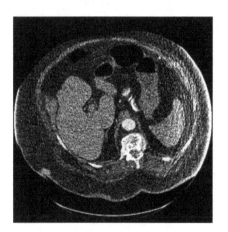

Fig. 2. A reconstructed WFBP image (left, center = 40 HU, width = 300 HU) and after high pass filtering (right).

$$I_{norm} = I - \bar{I} \qquad (3)$$

where I_{norm} is the normalized image obtained after high pass filtering (Fig. 2). The high pass filtering helps to focus on errors which are more apparent to the visual system and therefore affect subjective image reconstruction quality.

2.4   Convolutional Neural Network for Image Quality Score Estimation

Similar to Galdran et al. [2], we design a CNN to perform regression. Our CNN is trained on the noisy reconstructed CT images. Our aim is to learn a mapping from the noisy image to the GSSIM score. This would enable us to measure the reconstruction quality of a reconstructed image without requiring a reference image. Our CNN has 8 convolutional layers. The first two convolutional layers have 16 filters, the next two have 32 filters, and the following layers have 64, 128, 256 and 512 filters. The kernel size is 3 × 3. Each convolutional layer has a leaky ReLU activation function [9] to prevent vanishing gradients. The output connects to a global average pooling layer, which then feeds into a fully connected layer to predict the Image Reconstruction Quality Metric (IRQM) score. The fully connected layer has 512 neurons and a leaky ReLU activation function. The second and fourth convolutional layers perform strided convolutions with a stride size of 2. The architecture can be seen in Fig. 3.

The L2 norm between the predicted IRQM and the real GSSIM score is used as a loss value. The loss L_e is given by the following equation:

$$L_e = \frac{1}{N} \sum_{i=1}^{N} (GSSIM_i - IRQM_i)^2 \tag{4}$$

where N is the number of images.

2.5   Heatmap Regression

We introduce the use of error maps as an intermediate target for estimating image quality. The error maps are calculated by calculating the absolute difference between the reduced dose images with the full dose images. The error map E is given by

$$E = |I - I_r| \tag{5}$$

where I is the full dose image and I_r is the reduced dose image. Because we want the network to learn a quality score based on errors caused due to noise, we introduce an auxiliary loss function. We generate a heatmap of our neural network using gradient saliency [15], and use the L1 norm of our generated heatmap H with our error map E as an auxiliary loss (Fig. 3). Our total loss L is given by:

$$L = L_e + \frac{1}{N} \sum_{i=1}^{N} |H_i - E_i| \tag{6}$$

where N is the number of images and L_e is the loss introduced in Eq. 4. The total loss L was used for training our network. The network is trained with a learning rate of 5×10^{-5} over 50 epochs. The loss gradients are calculated at each node of the network using the backpropagation algorithm [20]. The weights of the network are updated using an Adam optimizer [5]. The minibatch size is 8.

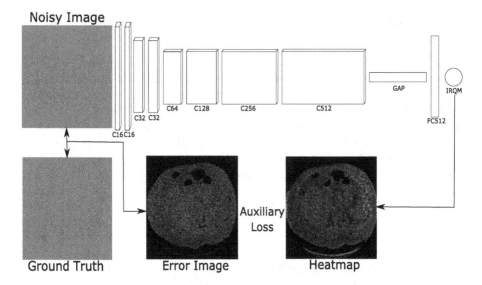

Fig. 3. Our CNN architecture to predict the IRQM score. C represents a convolutional layer with a respective number of filters, while FC represents a dense layer with respective number of neurons. The auxiliary loss generated with the heatmap and error map is also shown.

3   Experiments and Results

3.1   Comparison with Conventional Quality Metrics

The IRQM is compared with normal image quality metrics like the MSE and PSNR and perceptual metrics like SSIM and LPIPS [25]. We use two metrics to compare our performance to existing image quality metrics, the linear correlation coefficient (r), and Spearman's rank correlation coefficient (ρ). The obtained quality score is regressed with a five parameter logistic function [14], following which the r is calculated. The r values show a very strong correlation of IRQM with the SSIM ($r = 0.98$) and PSNR ($r = 0.98$), and a very strong inverse correlation with the LPIPS ($r = -0.91$) and MSE ($r = -0.96$) on the holdout dataset The r values show a very strong correlation of IRQM with the SSIM ($r = 0.92$) and PSNR ($r = 0.94$), and a very strong inverse correlation with the LPIPS ($r = -0.77$) and MSE ($r = -0.91$) on the test dataset The ρ values show a very strong correlation of IRQM with the SSIM ($\rho = 0.89$) and PSNR ($\rho = 0.95$) and very strong inverse correlations with the LPIPS ($\rho = -0.87$) and MSE ($\rho = -0.95$) on the holdout dataset The ρ values show a very strong correlation of IRQM with the SSIM ($\rho = 0.91$) and PSNR ($\rho = 0.94$) and very strong inverse correlations with the LPIPS ($\rho = -0.79$) and MSE ($\rho = -0.91$) on the test dataset Detailed results can be seen in Table 1 and Fig. 4.

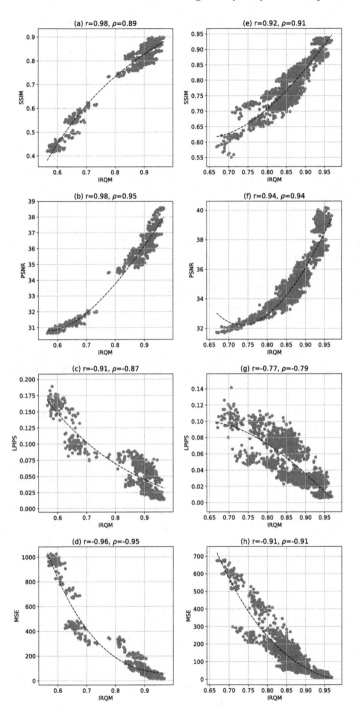

Fig. 4. r and ρ for the holdout set ((a)–(d)) and the test set ((e)–(h)) with the SSIM ((a), (e)) the PSNR ((b), (f)) the LPIPS ((c), (g)) and the MSE ((d), (h)). The black dashed lines show the line of best fit after non-linear regression.

Table 1. r and ρ values for the test and holdout datasets with all full reference metrics

Metric	Correlation coefficient (r)		Spearman coefficient (ρ)	
	Holdout data	Test data	Holdout data	Test data
SSIM	0.98	0.92	0.89	0.91
PSNR	0.98	0.94	0.95	0.94
LPIPS	−0.91	−0.77	−0.87	−0.79
MSE	−0.96	−0.91	−0.95	−0.91

3.2 Qualitative Results

Listwise Ranking Consistency: The effectiveness of the IRQM is tested using the listwise ranking consistency test [8]. We rank the IRQM with the radiation dose needed to produce them. An image reconstructed with a higher dose would be expected to have a higher IRQM score than that of an image reconstructed with a lower simulation dose. The ranking is assessed using Spearman's correlation coefficient (ρ) and Kendall's tau coefficient (τ). The mean value of ρ and τ are used to compare the performance. In order to avoid bias, the IRQM was tested on the data of a patient with 223 slices which was not present in the training, holdout, and test datasets. The patient data was reconstructed using WFBP and ADMIRE (strength 3) [12] with full, half and quarter doses. The IRQM achieved a mean ρ of 0.98 and a mean τ of 0.98.

Comparison with Iterative Reconstruction: We use the IRQM to compare the noise content of images reconstructed with simple WFBP and images reconstructed with iterative reconstruction. We reconstruct the data of the patient used in the Listwise Ranking Test with WFBP and ADMIRE (strength 3) with full dose, half dose, and quarter dose simulations. To compare noise levels at each dose, we use a two sample t-test. Significant improvements in noise level were observed at half dose ($t = -2.83$, $p = 0.00$) and full dose ($t = -12.09$, $p = 0.00$) (Fig. 5). No significant improvement in noise content was observed at quarter dose ($t = 0.03$, $p = 0.9$). This result was confirmed visually. The full dose WFBP reconstructions showed lower noise levels than half dose ADMIRE reconstructions ($t = 3.02$, $p = 0.00$).

Dose Reduction: We reconstructed the images of a single patient with different doses to see if the IRQM could adequately detect quality reduction in accordance with the dose reduction. The images were reconstructed with full dose, half dose, quarter dose, 10% dose, and 5% dose respectively. One way ANOVA was used to measure the differences in the measured IRQM values in different doses. A significant difference was observed among the population means ($F = 121.12$, $p = 0.00$). Moreover, Fig. 5 shows that the population means are significantly lower with dose reduction. However, images reconstructed at 10% dose are not significantly poorer quality than images reconstructed with quarter dose.

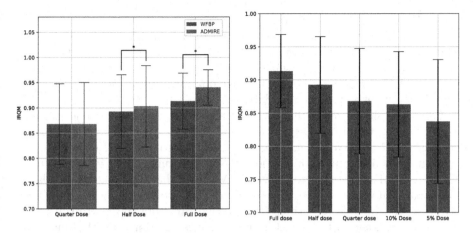

Fig. 5. Left is a bar plot representing the means of the reconstructed images with WFBP and ADMIRE algorithms. The blue bars represent WFBP and the red bars represent ADMIRE. The black bars represent error bars with 95% confidence intervals. Significant differences are denoted by asterisks. The differences are assessed using two-sample t-tests. For half dose reconstructions, the t value was -2.83 and the p value was 0.00, indicating a statistically significant difference between the two populations. For full dose reconstructions, the t value was -12.09 and the p value was 0.00, indicating a statistically significant difference between the two populations. Right is bar plot showing the means of the IRQM values for reconstructed images with various doses. The black bars represent error bars with 95% confidence intervals. The measured quality is lower with reduced dose. Differences between the populations were assessed using one-way ANOVA. The F value was 121.12 and the p value was 0.00, indicating a statistically significant difference between the populations. (Color figure online)

4 Discussion

In this study, we have investigated the possibility of using deep learning to quantify quantum noise in CT images. The developed IRQM showed a high correlation with existing ground truth metrics. In image quality assessment, a good relationship between a metric and image quality should be monotonic, but not necessarily linear. This is because, image fidelity typically decreases monotonically with increase in noise levels [8]. The IRQM has strong monotonic relationships with existing image quality metrics, which is indicated by the high ρ values. The relationship is also strongly linearly correlated following nonlinear regression, as evidenced by the high r values. This indicates that the IRQM can be used as a no-reference quality metric for CT images.

Our network accurately learned the noise levels present in CT images. This is proved by the listwise ranking test, where the IRQM ranked the images according to the simulated dose levels accurately. The IRQM was able to quantify noise from iterative reconstruction algorithms, which were not part of the original training dataset The IRQM showed that ADMIRE reconstructions showed significant improvements in noise content compared to WFBP at higher doses.

At low doses (75% dose reduction), no significant improvement was seen using the iterative reconstruction methods. Our results are in concordance with Winklehner et al. [21], which demonstrated a dose reduction of 50% in body CT without loss of image quality, while using iterative reconstruction. The IRQM was also able to learn the decrease in image quality even in cases of doses lower than those present in the training dataset.

A final consideration is the potential of our network in machine learning based CT image denoising. Deep learning based strategies have been used in CT denoising [7,22–24]. Wolterink et al. [22] used a mean squared error function for accuracy which resulted in the loss of small features. Yang et al. [24] used VGG-Net as a feature extractor, however, VGG-Net has been trained on a large image space and may not be able to learn CT noise features accurately. Our network would be able to extract features that are more specific to CT noise, giving an advantage in terms of CT image denoising. Moreover, the small size and simple architecture of our network compared to VGG-Net makes it a better choice for inclusion in CT reconstruction pipelines.

5   Conclusion

Our work has shown the possibility of developing a no-reference quality metric for measuring the amount of quantum noise present in a CT image without a reference image. The IRQM showed the statistical improvement in noise content when using iterative reconstruction at higher doses, as well as the decrease in image quality with dose reduction. Furthermore, the IRQM correlates well with other CT quality metrics when a ground truth image is present. The network trained could be used to assess and guide the training of deep learning based methods for CT image denoising.

References

1. Chen, G.H., Yang, C.L., Xie, S.L.: Gradient-based structural similarity for image quality assessment. In: International Conference on Image Processing, pp. 2929–2932, no. 1 (2006)
2. Galdran, A., Costa, P., Bria, A., Araujo, T., Mendonca, A.M., Campilho, A.: A no-reference quality metric for retinal vessel tree segmentation. In: Proceedings of the International Conference on Medical Image Computing and Computer Assisted Intervention, vol. 1, pp. 82–90 (2018). https://doi.org/10.1007/978-3-030-00928-1
3. Kijewski, M.F., Judy, P.F.: The noise spectrum of CT images. Phys. Med. Biol. **32**, 565–575 (1987)
4. Kim, J., Nguyen, A.D., Lee, S.: Deep CNN-based blind image quality predictor. IEEE Trans. Neural Netw. Learn. Syst. **30**(1), 11–24 (2019). https://doi.org/10.1109/TNNLS.2018.2829819
5. Kingma, D.P., Bai, J.L.: Adam: a method for stochastic optimization. In: International Conference on Learning Representations (2015). https://doi.org/10.1063/1.4902458

6. Li, C., Bovik, A.C.: Content-partitioned structural similarity index for image quality assessment. Signal Process. Image Commun. **25**(7), 517–526 (2010). https://doi.org/10.1016/j.image.2010.03.004

7. Liu, P., Li, Y., El Basha, M.D., Fang, R.: Neural network evolution using expedited genetic algorithm for medical image denoising. In: Frangi, A.F., Schnabel, J.A., Davatzikos, C., Alberola-López, C., Fichtinger, G. (eds.) MICCAI 2018. LNCS, vol. 11070, pp. 12–20. Springer, Cham (2018). https://doi.org/10.1007/978-3-030-00928-1_2

8. Ma, K., et al.: Waterloo exploration database: new challenges for image quality assessment models. IEEE Trans. Image Process. **26**(2), 1004–1016 (2017). https://doi.org/10.1109/TIP.2016.2631888

9. Maas, A.L., Hannun, A.Y., Ng, A.Y.: Rectifier nonlinearities improve neural network acoustic models. In: International Conference on Machine Learning, vol. 30, p. 6 (2013)

10. Maier, A., Fahrig, R.: GPU denoising for computed tomography. In: Xun, J., Jiang, S. (eds.) Graphics Processing Unit-Based High Performance Computing in Radiation Therapy, 1st edn, pp. 113–128. CRC Press, Boca Raton (2015)

11. Oppelt, A.: Noise in computed tomography. In: Aktiengesselschaft, S. (ed.) Imaging Systems for Medical Diagnostics, 2 edn., chap. 13.1.4.2, p. 996. Publicis Corporate Publishing (2005). https://doi.org/10.1145/2505515.2507827

12. Ramirez-Giraldo, J.C., Grant, K.L., Raupach, R.: ADMIRE: Advanced Modeled Iterative Reconstruction (2015). https://www.siemens-healthineers.com/computed-tomography/technologies-innovations/admire

13. Renieblas, G.P., del Castillo, E.G., Gómez-Leon, N., González, A.M., Nogués, A.T.: Structural similarity index family for image quality assessment in radiological images. J. Med. Imaging **4**(3), 035501-1–11 (2017). https://doi.org/10.1117/1.jmi.4.3.035501

14. Sheikh, H.R., Sabir, M.F., Bovik, A.C.: A statistical evaluation of recent full reference image quality assessment algorithms. IEEE Trans. Image Process. **15**(11), 3441–3452 (2006). https://doi.org/10.1109/TPCG.2004.1314471

15. Simonyan, K., Vedaldi, A., Zisserman, A.: Deep inside convolutional networks: visualising image classification models and saliency maps. In: International Conference on Learning Representations, pp. 1–8 (2013)

16. Stierstorfer, K., Rauscher, A., Boese, J., Bruder, H., Schaller, S., Flohr, T.: Weighted FBP - a simple approximated 3D FBP algorithm for multislice spiral CT with good dose usage for arbitrary pitch. Phys. Med. Biol. **49**(11), 2209–2218 (2004). https://doi.org/10.1088/0031-9155/49/11/007

17. Verdun, F.R., et al.: Image quality in CT: from physical measurements to model observers. Physica Medica **31**(8), 823–843 (2015). https://doi.org/10.1016/j.ejmp.2015.08.007

18. Wang, Z., Bovik, A.C., Sheikh, H.R., Member, S., Simoncelli, E.P., Member, S.: Image quality assessment: from error visibility to structural similarity. IEEE Trans. Image Process. **13**(4), 600–612 (2004)

19. Wang, Z., Simoncelli, E.P., Bovik, A.C.: Multiscale structural similarity for image quality assessment. In: The Thrity-Seventh Asilomar Conference on Signals, Systems & Computers, vol. 2, pp. 1398–1402 (2003). https://doi.org/10.1042/BJ20071051

20. Werbos, P.: Beyond regression: new tools for prediction and analysis in the behavioral sciences. Ph.D. thesis, Harvard University (1974). https://doi.org/10.1.1.41.8085

21. Winklehner, A., et al.: Raw data-based iterative reconstruction in body CTA: evaluation of radiation dose saving potential. Eur. Radiol. **21**(12), 2521–2526 (2011). https://doi.org/10.1007/s00330-011-2227-y
22. Wolterink, J.M., Leiner, T., Viergever, M.A., Isgum, I.: Generative adversarial networks for noise reduction in low-dose CT. IEEE Trans. Med. Imaging **36**(12), 2536–2545 (2017). https://doi.org/10.1109/TMI.2017.2708987
23. Xu, Q., Zhang, L., Yu, H., Mou, X., Hsieh, J., Wang, G.: Low-dose X-ray CT reconstruction via dictionary learning. IEEE Trans. Med. Imaging **9**(31), 1682–1697 (2012). https://doi.org/10.1016/j.pmrj.2014.02.014.Lumbar
24. Yang, Q., et al.: Low-dose CT image denoising using a generative adversarial network With Wasserstein distance and perceptual loss. IEEE Trans. Med. Imaging **37**(6), 1348–1357 (2018). https://doi.org/10.1109/TMI.2018.2827462
25. Zhang, R., Isola, P., Efros, A.A., Shechtman, E., Wang, O.: The unreasonable effectiveness of deep features as a perceptual metric. In: Proceedings of the IEEE Computer Society Conference on Computer Vision and Pattern Recognition, pp. 586–595, no. 1 (2018). https://doi.org/10.1109/CVPR.2018.00068

Deep Learning Based Metal Inpainting in the Projection Domain: Initial Results

Tristan M. Gottschalk[1,2(✉)], Björn W. Kreher[3], Holger Kunze[3], and Andreas Maier[1,2]

[1] Friedrich-Alexander-University Erlangen-Nuernberg, Erlangen, Germany
tristan.gottschalk@fau.de
[2] Erlangen Graduate School in Advanced Optical Technologies (SAOT), Erlangen, Germany
[3] Advanced Therapies, Siemens Healthcare GmbH, Forchheim, Germany

Abstract. During surgical interventions mobile C-arm systems are used in order to evaluate the correct positioning of e.g. inserted implants or screws. Besides 2D X-ray projections, that often do not suffice for a profound evaluation, new C-arm systems provide 3D reconstructions as additional source of information. However, mainly due to metal artifacts, this additional information is limited. Thus, metal artifact reduction methods were developed to resolve these problems, but no generally accepted approaches have been found yet. In this paper, three different network architectures are presented and compared that perform an inpainting of metal corrupted areas in the projection domain in order to tackle the problems of metal artifacts in the 3D reconstructions. All network architectures were trained using real data and thus all observations should hold during inference in real clinical applications. The network architectures show promising inpainting results with smooth transitions with the non-metal areas of the images and thus homogeneous image impressions. Furthermore, this paper shows that providing additional input data to the network, in form of a metal mask, increases the inpainting performance significantly.

Keywords: Metal artifact reduction · X-ray · C-arm · Inpainting

1 Introduction

Besides the permanently installed imaging modalities like Magnetic Resonance Imaging and Computed Tomography, mobile X-ray modalities like C-arms exist. Modern C-arms are not only capable of acquiring 2D X-ray projections, but also 3D reconstruction. One of the major areas of application of 3D reconstructions performed by a C-arm is the evaluation of correct positioning of e.g. inserted implants or screws during surgical interventions, because a solely evaluation using 2D projections often does not suffice. Although the acquired 3D volumes provide additional information during surgery, different image artifacts in the 3D

© Springer Nature Switzerland AG 2019
F. Knoll et al. (Eds.): MLMIR 2019, LNCS 11905, pp. 125–136, 2019.
https://doi.org/10.1007/978-3-030-33843-5_12

volume limit the benefit - especially metal artifacts, caused by the implants, are the restricting factor. Thus, metal artifact reduction (MAR) methods were developed in order to resolve these problems, but no generally accepted approaches have been found yet. Apart from classic MAR algorithms [1–4], particularly deep learning based methods show promising results. Whereas some of the methods try to tackle the metal artifacts in image-domain [5,6] by learning some kind of destreaking, most approaches process the data in projection domain [7–9]. Typically, in these methods, the so called metal trace is removed from the acquired sinogram data and the missing information is subsequently inpainted using different network architectures. A sole handling of metal artifacts in image-domain as presented by Gjesteby et al. [5] and Huang et al. [6] might not be sufficient because these methods only reduce the artifacts in the reconstructed volumes but do not tackle the underlying problems of inconsistencies caused by the metal corrupted projections. Due to the fact that the representation of projection data as sinograms only hold for the central slice in cone-beam geometry, a metal inpainting network is proposed that directly uses the acquired 2D projections as input. Furthermore, Unberath et al. [14] were able to present promising results in an inpainting task for Virtual DSA using 2D projections. Additionally, the representation of the input data as projections might simplify the inpainting task since it is possible to visually perceive underlying structures as e.g. bone edges through the metal objects.

2 Proposed Method

In the following section, three different implementations of metal inpainting network architectures with increasing complexity are described in detail. Furthermore, the acquisition of corresponding metal corrupted and metal free projections is explained. In addition to that, the section covers the performed preprocessing, the split into train, validation and test data followed by a detailed overview about the training procedure.

2.1   Network Architectures

All network architectures are based on the U-Net [10] structure and use one common basic concept – a network with eight contracting and eight expanding blocks. Each of the contracting blocks consists of two convolutional layers using 3×3 kernels and Rectified Linear Units (ReLU) as activation function, followed by a max-pooling layer using a 2×2 kernel. In contrast, the blocks in the expanding part of the network consist of an upsampling layer followed by two convolutional layers, again using 3×3 kernels and ReLUs as activation function. As a last layer of the network, one additional convolution using a 1×1 kernel is implemented outputting the single-channel grayscale image. Starting with an input image of size $976 \times 976 \times 1$ and 8 feature maps per layer in the first block, the image size halves and the amount of feature maps doubles each block inside the contraction part of the network. This leads to a bottleneck with an image size of 7×7 pixels and 1024 feature maps. In the corresponding expansion part of the network, the feature maps halves and the image size doubles at each block, ending up with an output image with the original size of $976 \times 976 \times 1$. Additionally, each contracting block is wired with its corresponding expanding block using skip connections and concatenations. In contrast to the original U-Net architecture, one additional skip connection is implemented such that the network's input is directly concatenated with the output of the last expansion block before being solely processed in the last convolutional layer using a 1×1 kernel. That added skip connection should help the network to simplify the decision of using the input's original data or the processed output of the network's expanding part.

Using the explained basic architecture, the first and simplest network, *S-MAR-Net*, solely takes the corrupted metal projection as input and outputs a single inpainted projection. In contrast to that, the second network, *Mask-MAR-Net*, is additionally provided with a corresponding metal mask of the corrupted metal projection and has again a single inpainted projection as an output. The mentioned provided metal mask needed to be created in a preliminary step. In the course of this project, and in order to train the Mask-MAR-Net, the mask is simply acquired by subtracting the metal free ground truth projection from the metal corrupted projection and subsequently performing a thresholding. Thus, the network can only be used in inference if a preliminary processing step, as e.g. a segmentation network, provides a suitable metal mask. In order to deal with that problem, the third and last network architecture was developed, *Dual-MAR-Net*, which is a two stage architecture combining the two aforementioned networks. Dual-MAR-Net uses the implicitly learned metal segmentation of S-MAR-Net by subtracting S-MAR-Net's output from the input and using that as additional mask input for the Mask-MAR-Net. Thus, Dual-MAR-Net enables the usage during inference without the necessity of a preliminary mask-generation process – S-MAR-Net acts as the mask-generator. The three different architectures are illustrated in detail in Fig. 1.

2.2   Training the Network

Data. In order to train the proposed Metal Inpainting Networks it is necessary to have access to data, that consists of matching X-ray projections with and without superimposed metal objects that corrupt the data. Typically, such kind of data can only be provided by simulating the superimposition of metal objects into originally metal free projections. This kind of simulations have major drawbacks. Either the simulation physically correctly models the imaging process and the superimposition and thus achieve realistic data as e.g. a Monte Carlo Simulation but with the drawback of very high computational effort, or they are fast in generating data but with the drawback of a physically insufficient modeled image acquisition process. Thus, there is always a trade-off between reality and computational effort. Contrary to most MAR publications, these networks are trained using real data. For this paper, twelve corresponding 3D X-ray Cone-Beam scans of human knee cadavers with and without randomly placed metal

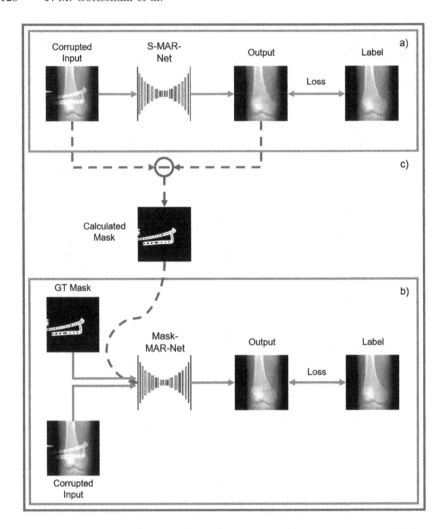

Fig. 1. Overview about the data flow of the three different network architectures. (a) illustrates the S-MAR-Net, (b) the Mask-MAR-Net and (c) the Dual-MAR-Net. The red dotted lines represents the data flow during inference using the Dual-MAR-Network. Instead of using the ground truth metal mask as additional input for the Mask-MAR-Net, the metal mask is calculated from the input and output of S-MAR-Net.

objects on the surface of the respective knees, were acquired. These 3D scans were performed using a Siemens Cios Spin C-arm system with a detector size of 30×30 cm. Thus, the resulting volumes cover a cube with edge length of 160 mm. Each of these 3D scans consists of 400 two dimensional X-ray projections with a size of 976×976 pixels and covering an angular range of a short scan, thus $180°$ plus the C-arm's fan angle. The correspondence between the metal and metal-free projections were achieved by performing two immediately consecutive scans. Figure 4(a) and (b) show one set of these corresponding 2D projections. During the two consecutive scans the metal objects were removed in a way such that there was no movement of the knee as well as of the scanner. Due to the fact that the acquisition of single projections during each 3D scan are not solely triggered by the current acquisition angle but rather by a number of parameters like e.g. the state of the detector, not every set of corresponding projections fit perfectly as a result of slightly different acquisition angles. Thus, a minority of corresponding projections show a tiny misalignment. This small misalignment could potentially aggravate the learning task for the network.

Preprocessing. The measured intensities at the detector can be converted into the line integral data by using the Lambert-Beer Law [13]. Thus, firstly, the measured intensities (I) are normalized by the initial intensity (I_0) submitted from the X-ray source. After that normalization, all values should theoretically lie in the range between 0 and 1. However, due to effects like e.g. X-ray scatter, especially at the metal objects, the actual values can slightly vary from that. With that normalization, the integral over the attenuation coefficient μ along the path x on the X-ray beam is calculated as follows:

$$\int \mu(x)dx = -\ln\left(\frac{I}{I_0}\right). \tag{1}$$

Train-Validation-Test Split. The twelve acquired 3D scans were split into eight scans for training, two scans for validation and two scans for testing, resulting in a training set with 3200 projections and a validation and a test set with 800 projections each.

Training Procedure. Using the eight preprocessed 3D scans, both, the S-MAR-Net as well as the Mask-MAR-Net, were separately trained using the Adam optimizer [11] with an exponentially decaying learning rate, starting with $1e^{-4}$, using mean squared error

$$L_{MSE} = ||y - \hat{y}||_2^2, \tag{2}$$

as loss function, where $\hat{y}$ denotes the network's output and y the ground truth label. Each of the network architectures were trained for 1000 Epochs using a batch size of 1.

3 Results

As explained previously, the different networks were trained and tested using real data. Thus, the results could be evaluated with full physical realism and consequently all observations should hold during inference in real clinical applications.

In addition to the visual evaluation of the results, the quantitative evaluation of the metal inpainting results of the different architectures and loss functions uses the Mean Squared Error (MSE) and the Peak Signal-to-Noise Ratio (PSNR). Both metrics are solely calculated in the metal corrupted areas of the image in order to provide a meaningful evaluation of the performed inpainting. Additionally to the inpainting results, the implicitly learned metal segmentation of the S-MAR-Net is evaluated using Intersection over Union (IoU) and Accuracy (Acc) in order to make the inpainting results more comparable.

3.1 Implicitly Learned Segmentation

Before the inpainting quality of the Dual-MAR-Net during inference can be evaluated, it is necessary to determine the capabilities of the implicitly learned segmentation of S-MAR-Net, because during inference the extracted mask serves as additional input for the subsequent Mask-MAR-Net (c.f. Fig. 1), thus forming the Dual-MAR-Net.

Depicted in Table 1, it can be seen that the two 3D test scans (each containing 400 projections) result in significantly different results. For the first test scan, the average IoU lies at 0.9738, whereas of the second test scan at 0.9312. Also the average accuracy shows a higher value of 0.9980 for the first test scan. The most significant difference between the scans can be seen in the minimal IoU raging between 0.8547 for the first and 0.5131 for the second test scan. Combining both test scans, an average IoU of 0.9525 and an average accuracy of 0.9962 is reached. Figure 2 shows two implicitly learned segmentation results.

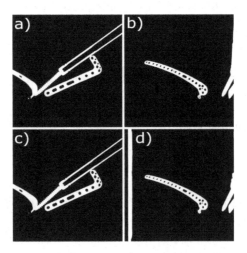

Fig. 2. Two examples of the implicitly learned segmentation of S-MAR-Net. (a) and (b) denotes the respective ground truth mask and (c) and (d) the learned segmentations. It can be seen that in (d) the examination table is wrongly segmented as well.

Table 1. IoUs and Accs of the implicitly learned segmentation for the complete test dataset and for each test scans separately.

Dataset	Min IoU	Max IoU	Avg. IoU	Min Acc	Max Acc	Avg. Acc
Test Data 1 + 2	0.5131	0.9947	0.9525	0.9524	0.9997	0.9962
Test Data 1	0.8547	0.9947	0.9738	0.9910	0.9997	0.9980
Test Data 2	0.5131	0.9817	0.9312	0.9524	0.9984	0.9945

3.2 Inpainting Results

Comparing the different network architectures by having a look at Fig. 3, shows that providing additional mask information to the Mask-MAR-Net did not significantly decrease the loss on the training dataset. The training losses of the S-MAR-Net and Mask-MAR-Net are almost identical with $1.8e^{-4}$ and $2.1e^{-4}$. Comparing the validation losses, Fig. 3 shows that the additional mask information improves the performance of the network from $3.0e^{-3}$ to $1.6e^{-3}$ by around $1.3e^{-3}$, when processing the "unseen" validation data. The quantitative evaluation in Table 2 shows that Mask-MAR-Net's average MSE of 0.0025 lie significantly below and its average PSNR of 31.129 dB significantly above those of the S-MAR-Net when evaluating using the whole test dataset. The average PSNR rises by roughly 4 dB and the MSE drops by 0.007 compared to S-MAR-Net. Having a look at the results of the second test scan, it becomes clear, that the performances of both networks, S-MAR-Net and Mask-MAR-Net lie significantly below that of the first test scan with average MSEs of 0.0154 and 0.0032 and average PSNRs of 24.419 dB and 30.109 dB. Taking a look at Fig. 4, the areas marked by the yellow arrows in the zoomed images, show that Mask-MAR-Net

Table 2. Quantitative evaluation of the different network architectures using the complete test dataset (TS 1 + 2) and for each of the two test scans separately (TS 1 and TS 2).

Dataset	Network	MSE			PSNR		
		Min	Max	Avg	Min	Max	Avg
TS1	S-MAR-Net	0.0012	0.0110	0.0033	23.3000	35.0000	29.4453
	Mask-MAR-Net	0.0007	0.0050	0.0017	26.6000	35.3000	32.1488
	Dual-MAR-Net	0.0011	0.0853	0.0041	17.1000	37.3000	29.5855
TS2	S-MAR-Net	0.0025	0.1100	0.0154	13.9000	33.7000	24.4192
	Mask-MAR-Net	0.0012	0.0129	0.0032	23.1000	34.7000	30.1092
	Dual-MAR-Net	0.0019	0.8750	0.0217	5.2400	31.7000	24.9920
TS 1 + 2	S-MAR-Net	0.0012	0.1100	0.0094	13.9000	35.0000	26.9289
	Mask-MAR-Net	0.0007	0.0129	0.0025	23.1000	35.3000	31.1289
	Dual-MAR-Net	0.0011	0.8750	0.0129	5.2400	37.3000	27.2888

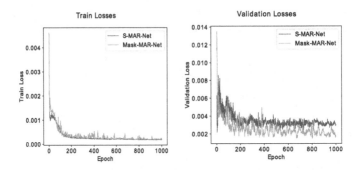

Fig. 3. Train and Validation Losses during training of the two different network architectures using L_{MSE}.

is able to restore the bone edges, whereas S-MAR-Net leaves a blurry edge. Furthermore, it is visible that the inpainting of the Mask-MAR-Net results in a better transition with the neighboring tissues and produces an overall more homogeneous image impression. Having a look at the full images in Fig. 4 shows that a slightly incorrect metal mask made the Mask-MAR-Net ignore metal corrupted pixels, whereas the implicitly learned metal segmentation of S-MAR-Net correctly selected all metal corrupted pixels.

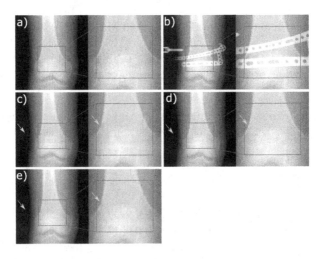

Fig. 4. Inpainting results of S-MAR-Net (c), Mask-MAR-Net (d) Dual-MAR-Net (e) and the corresponding input (b) and label (a) projections.

The inpainting results of the inference model Dual-MAR-Net also vary significantly over the two single 3D test scans. Processing the first test scan results in an average MSE of 0.0041 and an average PSNR of 29.586, whereas processing the second test scan results in an average MSE of 0.0217 and an average PSNR of 24.991. Thus, the inpainting quality that is reached evaluating with the second test scan is significantly worse. It is visible that the maximal MSE of the second test scan is 10 times higher then the one of the first test scan. Furthermore, it is significant that the range between the maximal and minimal MSE is wide in comparison to the one of the first test scan. Processing both test scans result in an average MSE of 0.0129 and an average PSNR of 27.288. Comparing the quantitative results with those of S-MAR-Net and Mask-MAR-Net it can be noted that Dual-MAR-Net outputs slightly higher PSNRs but lower MSEs as S-MAR-Net and inferior PSNRs and MSEs in comparison with Mask-MAR-Net. Having a look at Fig. 4(e) shows that Dual-MAR-Net provides comparable visual results to MASK-MAR-Net, when processing the first test set. It can be seen that Dual-MAR-Net outputs an inpainting result with a smooth transition to the neighboring tissue and thus generates an overall homogeneous image impression. However, the edges are a little bit more smoothed in comparison to Mask-MAR-Net.

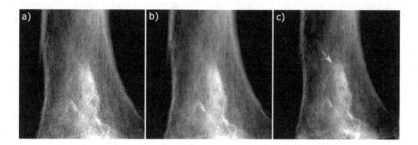

Fig. 5. Comparison between ground truth label (a), Dual-MAR-Net's output when processing label as input (b) and Dual-MAR-Net's output when processing metal corrupted projection as input (c). It can be seen that the processed images are blurred. The yellow arrow marks an area where inpainting was performed. These parts are blurrier than the rest of the image. (Color figure online)

Besides solely evaluating the performance of the inpainting task, especially in the field of medical data processing, it is very important to perform sanity checks with the network. One important sanity check is to evaluate what happens, when you feed a metal-free and thus non-corrupted image to the network. Ideally, the data should not be changed at all. Feeding the ground truth label projections of the first scan of the test set through Dual-MAR-Net showed that no critical changes were done to the label images – this also holds for the two other architectures. However, it became clear that the output of the Dual-MAR-Net seems to be slightly blurry, which can be seen in Fig. 5. Further comparing the output of a processed label projection with the output of a real metal corrupted projection, it can be noted that the inpainted areas of the output are again slightly blurrier than the rest of the image.

4   Discussion

The presented results regarding the implicitly learned segmentation show several interesting aspects. The wide range between the maximal and minimal IoU but the rather high average for the second test scan show that the segmentation seems to fail for only a few projections of the scan. Having a closer look at the implicitly generated metal masks, clarifies that in roughly the first 20 and last 20 projections the examination table is incorrectly segmented as metal, too. Apart from the wrongly added parts of the table, the segmentation of the metal objects works reliably (c.f. Fig. 2). Training the final network with a higher amount of more varying data in terms of different tables might fix this problem in the future. In contrast, the narrow range between those values in the first test scan, show that the segmentation works more stable and the network can clearly distinguish between the corrupting metal parts and the examination table. Based on the performance reached by testing with the first test scan and by keeping in mind that the inferior performance of the second test scan is due to an overestimation of the metal mask by wrongly including parts of the table, but correctly segmenting the metal, it can be stated that the intrinsically learned segmentation works on a comparable level with the ground truth mask in regards of finding all metal parts in the image. Thus, in the case of the first test scan, the differences in the results of the different network architectures should mainly be caused by the differences in the quality of the respective inpaintings and almost entirely independent from the segmentation.

The presented results regarding the different network architectures, show that the quality of the Mask-MAR-Net's as well as Dual-MAR-Net's inpainting highly correlates with the quality of the provided metal mask. Missing metal corrupted areas in the mask will lead to unprocessed metal corrupted pixels in the output image. Additionally, the visual and especially the quantitative results, show that providing additional information in form of a metal mask, significantly improves the inpainting performance on "unseen" data and thus during inference in the clinical application. This assumption is proven by the significantly higher quantitative values as well as by the more homogeneous image impression with simultaneously distinct bone edges of the Mask-MAR-Net. The high performance of the Mask-MAR-Net shows that it might be preferable to separate the segmentation task and inpainting task into two networks. Furthermore, the results have shown that S-MAR-Net and Dual-MAR-Net achieve comparable quantitative results but with Dual-MAR-Net achieving a visually more homogeneous image impression. In addition, it becomes clear that the results highly depend on the quality of the intrinsic segmentation of S-MAR-Net and thus on whether parts of the table are wrongly segmented as metal or not. The worse performance of Dual-MAR-Net in comparison with Mask-MAR-Net might be explained by the fact that Dual-MAR-Net was never trained jointly and thus its second network part (Mask-MAR-Net) has never seen other masks than the ground truth masks. This potentially leads to severe errors in cases of wrongly masking non metal areas like e.g. the table, because the network has never seen such tasks during training with the ground truth masks. As a result, a jointly retraining of Dual-MAR-Net might increase its performance significantly and might obviate the need of a preliminary masking process.

Lastly, the results of the sanity check show that the U-Net-like architecture is not able to completely restore the high frequencies of the input, although there is an additional skip connection linking the input to the last convolutional layer of the network. A possible solution for that might be a fusion of the inpainted parts (those inside the metal mask) of the network's output with the unprocessed original data from the input projections in all non-metal areas.

5 Conclusion

In this paper, we have compared three different metal inpainting network architectures working with 2D X-ray projection data. It is shown that providing a metal mask as additional input to the network significantly increases the inpainting performance visually as well as quantitatively. Thus, it might be beneficial to separate the segmentation and the inpainting tasks into two networks. As a further step, it is shown that a two stage network, called Dual-MAR-Net, could be developed that combines the two separately learned metal inpainting networks S-MAR-Net and Mask-MAR-Net in such a way that it reaches the quantitative performance of S-MAR-Net but with a visually more homogeneous image impression. Additionally, it is demonstrated that the proposed networks do not induce critical changes in the non-metal areas. However, the network architectures are not able to fully restore all high frequencies of the input image. As a result, it induces a slight blurring of the whole image. Thus, future work will include testing perceptual loss functions and a fusion of output and input image in order to improve the restoration of more high frequencies. In addition it will be tested to separate the segmentation and inpainting task into two networks. Furthermore, beyond the scope of this paper, a brief look at the corresponding 3D reconstructions using the inpainted projections has shown that information of neighboring projections or volume-domain specific information need to be included to the training process in order to solve inconsistencies between the inpainted areas of consecutive projections of the 3D scan in order to create artifact free reconstructions. This should be achieved by using the Python Reconstruction Operators in Neural Networks (PYRO-NN) [12].

Acknowledgments. This work was supported by the Siemens Healthcare GmbH, 91301 Forchheim, Germany.

References

1. Meyer, E., Raupach, R., Lell, M., Schmidt, B., Kachelrieß, M.: Frequency split metal artifact reduction (FSMAR) in computed tomography. Med. Phys. **39**(4), 1904–1916 (2012)
2. Meyer, E., Raupach, R., Lell, M., Schmidt, B., Kachelrieß, M.: Normalized metal artifact reduction (NMAR) in computed tomography. Med. Phys. **37**(10), 5482–5493 (2010)

3. Xinhui, D., Li, Z., Yongshun, X., Jianping, C., Zhiqiang, C., Yuxiang X.: Metal artifact reduction in CT images by sinogram TV inpainting. In: 2008 IEEE Nuclear Science Symposium Conference Record, pp. 4175–4177. IEEE, Dresden (2008)
4. Kalender, W.A., Hebel, R., Ebersberger, J.: Reduction of CT artifacts caused by metallic implants. Radiology **164**(2), 576–577 (1987)
5. Gjesteby, L., et al.: Deep learning methods for CT image-domain metal artifact reduction. In: Developments in X-Ray Tomography XI, vol. 10391, 103910W pages. International Society for Optics and Photonics (2017)
6. Huang, X., Wang, J., Tang, F., Zhong, T., Zhang, Y.: Metal artifact reduction on cervical CT images by deep residual learning. Biomed. Eng. Online **17**(1), 175 (2018)
7. Ghani, M.U., Clem Karl, W.: Deep learning based sinogram correction for metal artifact reduction. Electron. Imaging **2018**(01), 4721–4728 (2018)
8. Claus, B.E.H., Jin, Y., Gjesteby, L.A., Wang, G., De Man, B.: Metal-artifact reduction using deep-learning based Sinogram completion: initial results. In: Proceedings of 14th International Meeting Fully Three-Dimensional Image Reconstruction Radiol. Nucl. Med., pp. 631–634. Fully3D (2017)
9. Park, H.S., Lee, S.M., Kim, H.P., Seo, J.K., Chung, Y.E.: CT sinogram-consistency learning for metal-induced beam hardening correction. Med. Phys. **45**(12), 5376–5384 (2018)
10. Ronneberger, O., Fischer, P., Brox, T.: U-Net: convolutional networks for biomedical image segmentation. In: Navab, N., Hornegger, J., Wells, W.M., Frangi, A.F. (eds.) MICCAI 2015. LNCS, vol. 9351, pp. 234–241. Springer, Cham (2015). https://doi.org/10.1007/978-3-319-24574-4_28
11. Kingma, D. P., Ba, J.: Adam: A method for stochastic optimization. arXiv preprint arXiv:1412.6980 (2014)
12. Syben, C., Michen, M., Stimpel, B., Seitz, S., Ploner, S., Maier, A.K.: PYRO-NN: Python Reconstruction Operators in Neural Networks. arXiv preprint arXiv:1904.13342 (2019)
13. Maier, A., Steidl, S., Christlein, V., Hornegger, J.: Medical Imaging Systems. Springer, New York (2018). https://doi.org/10.1007/978-3-319-96520-8
14. Unberath, M., Hajek, J., Geimer, T., Schebesch, F., Amrehn, M., Maier, A.: Deep learning-based inpainting for virtual DSA. In: IEEE Nuclear Science Symposium and Medical Imaging Conference (2017)

Deep Learning for General Image Reconstruction

Flexible Conditional Image Generation of Missing Data with Learned Mental Maps

Benjamin Hou[✉], Athanasios Vlontzos, Amir Alansary, Daniel Rueckert, and Bernhard Kainz

Biomedical Image Analysis Group, Imperial College London, London, UK
bh1511@imperial.ac.uk

Abstract. Real-world settings often do not allow acquisition of high-resolution volumetric images for accurate morphological assessment and diagnostic. In clinical practice it is frequently common to acquire only sparse data (e.g. individual slices) for initial diagnostic decision making. Thereby, physicians rely on their prior knowledge (or mental maps) of the human anatomy to extrapolate the underlying 3D information. Accurate mental maps require years of anatomy training, which in the first instance relies on normative learning, i.e. excluding pathology. In this paper, we leverage Bayesian Deep Learning and environment mapping to generate full volumetric anatomy representations from none to a small, sparse set of slices. We evaluate proof of concept implementations based on Generative Query Networks (GQN) and Conditional BRUNO using abdominal CT and brain MRI as well as in a clinical application involving sparse, motion-corrupted MR acquisition for fetal imaging. Our approach allows to reconstruct 3D volumes from 1 to 4 tomographic slices, with a SSIM of 0.7+ and cross-correlation of 0.8+ compared to the 3D ground truth.

1 Introduction

Physical as well as physiological constraints on tomographic image acquisition (e.g. motion) often prohibit the acquisition of high resolution volumetric images that are commonly used for morphological examinations and diagnosis. Acquisition of high resolution images requires a fixed period of time where the patient is asked to remain still, this is often not possible in cases such as fetal imaging. Motion during this period causes scanned slices to become incoherent and corrupt. Long periods of CT scans also impose high levels of exposure to ionising radiation. Single slice or sparse acquisition can often be mentally extrapolated to a 3D mental map by experienced physicians. However, it relies on years of experience and training, thus the need to perform sparse reconstruction arises. In this paper, we address both the need to perform sparse reconstruction, as well as creating mental maps of anatomies.

Extrapolation of 3D volumes have advantages for tracking and interventional applications. Tracking, e.g. methods such as freehand ultrasound, can benefit the sonographer greatly by providing an extrapolated 3D volume for better spatial reference. Furthermore, iterative image-based motion compensation methods

© Springer Nature Switzerland AG 2019
F. Knoll et al. (Eds.): MLMIR 2019, LNCS 11905, pp. 139–150, 2019.
https://doi.org/10.1007/978-3-030-33843-5_13

needs a good initial target, which is often not possible to obtain if the subject is awake and constantly in motion during image acquisition. Thus, the need to extrapolate a full 3D volume from very sparse amount of slices is highly desirable.

We leverage Bayesian Deep Learning (BDL) and environment mapping to generate full volumetric anatomy representations derived from none to a few conditioning slices. In contrast to commonly used Conditional Variational AutoEncoders (C-VAE), our model leverages traditional statistical methods where the conditioning variable is not fixed or restricted. This therefore enables us to perform reconstruction of normative structures, extrapolate sparse image acquisition and create mental maps of anatomies. Contrary to previous approaches of sparse reconstruction, as detailed in the related work section below, our method can also produce probabilistic mental representations of the anatomy and anatomical context in question to aid diagnosis and therapy.

Related Work: Sparse reconstruction of anatomical structures has been the topic of extensive work as a method to reduce cost and, *e.g.*, exposure to ionising radiation for patients and doctors alike. Early approaches included deformable statistical models [5] to set a prior to the reconstruction process. More recent approaches have been adopting neural networks and deep learning to perform sparse reconstruction. Cerrolaza et al. [1] uses a hierarchical C-VAE, where given three standard plane views from a 2D ultrasound scan of a fetal brain, to reconstruct the 3D segmentation mask of the fetal skull. Similarly [17] use a Convolutional Auto-Encoder to construct a shared latent space between 2D and 3D images to aid the reconstruction of a 3D image. In addition [3] perform an inter-domain sparse reconstruction as they perform segmentation of 3D volumes based on 2D sparse data inputs. In the field of natural images [2] suggested an iterative technique of refining the 3D reconstruction as the model is given more views. Finally in [12] used stereoscopic reconstruction to achieve 2D to 3D segmentation reconstruction.

Contribution: We introduce a method to generate missing slices, via BDL, by sampling from a distribution on the image manifold, which is conditioned on sparse scanned slices as context. We restructure the Conditional BRUNO [11] architecture, and train the model to learn a mental map of a specific region of interest or anatomy. The novel aspect of our work is that conditional image generation is not achieved by commonly used C-VAE architectures, but instead through Normalising Flows [14] and statistical modelling such as Student's t-process. This is applied to generate patient specific dense medical volumes, and evaluated on three different data-sets. To generate patient specific dense medical volumes, we query the model by performing a dense sweep of all possible pose positions, while conditioning the model on sparsely sampled context slices from the patient. The method is evaluated on three data-sets and we demonstrate its application for motion correction in fetal brain MRI.

Background: Generative Models are used to model probabilistic distributions, $p(x)$, of a data-set, X, such that $x \in X$ in some high-dimensional space $\mathcal{X}$. The model can then be used to generate new samples, such as images, that follows the same probabilistic distribution. New samples are seeded by a latent variable, a vector often denoted z in some high-dimensional space $\mathcal{Z}$, and are sampled according to some Probability Density Function (PDF) $p(z)$. Given a fixed deterministic function, $f(z; \theta)$, parameterised by θ in some high-dimensional space Θ and $f : \mathcal{Z} \times \Theta \to \mathcal{X}$, the aim is to optimise θ such that samples of z from $p(z)$, and subsequently $f(z; \theta)$, will be similar to x with high probability. Formally, this can then be written as: $p(x) = \int p_\theta(x|z) p_\theta(z) dz$.

The most commonly used distribution, also known as the prior, for the latent space is a Gaussian, with a mean of zero and unit variance ($\mathcal{N}(0, 1)$). An important component of the modelling process is defining a bijector, which is an invertible fixed transformation function that maps one data space to another. f, defined above, can be used to map the complex distribution of the input data space to the z latent space. The distribution of modelled image space can then be written as: $X \sim \mathcal{N}(f(z; \theta), \sigma^2 I)$. As the probability integral is high-dimensional and complex, a neural network can be used to learn f. Therefore it is possible to use gradient descent (or any other optimisation technique) to perform $\max_\theta \sum_i \log(p_\theta(x_i))$, which aims to find an optimal set of parameters θ, for the fixed deterministic function $f(z; \theta)$.

Such mappings can be achieved through methods such as Variational AutoEncoders (VAEs) [10] or Generative Adversarial Networks (GANs) [7]. The encoder and decoder component of the VAE models the forward and inverse of the bijector function, but are learned separately. In GANs, only the fixed function from latent z space to data space is learned. RealNVP [4] and Masked Autoregressive Flow (MAF) [13], however, are bijectors that are fully invertible. The same weights are used for both forward and inverse transformations.

Conditional Generative Models, such as C-VAEs [16], Generative Query Network (GQN) [6], or BRUNO [11], generate new samples based on a predefined condition such that for each possible value of c there exists a $p(z)$; $p(x|c) = \int p_\theta(x|z) p_\theta(z, c) dz$. For this particular task, c is a set of images that are sparsely acquired, and is not bound by quantity. This requires the set of images in the condition being exchangeable, i.e. the joint probability is invariant to permutation of the images. For any permutation, π; $p(x_1, x_2, ..., x_n) = p(x_{\pi_1}, x_{\pi_2}, ..., x_{\pi_n})$. Random variables are often independent and identically distributed (iid), and iid random variables are always infinitely exchangeable. However, the converse is not always true, an infinitely exchangeable sequence is not necessarily iid. Bruno de Finetti's theorem therefore states 'a sequence of random variables $(x_1, x_2, ..., x_n)$ is infinitely exchangeable *iff* for all n; $p(x_1, x_2, ..., x_n) = \int \prod_{i=1}^{n} p(x_i|\theta) p(\theta) d\theta$. The stochastic process is then defined; $p(x_n|x_{1:n-1}) = \int p_\theta(x_n|z) p_\theta(z|x_{1:n-1}) dz$.

2   Method

In the proposed framework, the Conditional BRUNO architecture is restructured and trained to build mental maps of medical volumes using 2D slices, x, with their corresponding pose parameter, v, that represents the slice's location in 3D space. Similar to a GQN, the data-set is of the form $D = \{(x_n^k, v_n^k)\}$, where

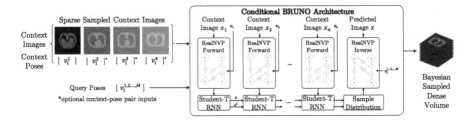

Fig. 1. BRUNO Architecture for the generation of volumetric anatomical mental maps from very sparse data. Conditioning context images $v_i^{1,2,...,M}$ can be any sample from the observed anatomy in contrast to defined samples in, e.g., C-VAEs.

$n \in \{1,2,...,N\}$ and $k \in \{1,2,...,K\}$. N is the number of high resolution 3D volumes and K is the number of 2D axial slices of the volumes. During training, M random image-pose pairs, are sampled from a particular volume. Each $m \in M$ is a particular observation, with the collective being denoted as a **sequence**. $M-1$ observations from the sequence are used as **contexts**, with the remaining image-pose pair being used as the **query pose** and **target image** (Fig. 1).

Each individual context (i.e. image-pose pair) in a sequence is passed through a Conditional RealNVP [4], which is the bijector of the model. The affine coupling layer uses Convolutional Neural Networks (CNN), and learns the mapping of the input image distribution to a Gaussian prior. The mapping is conditioned on the pose, and is made possible by augmenting the input image with the pose vector as an additional input variable. This CNN can be of any structure, for simplicity, a simplified ResNet was used. As the RealNVP component is trained on a Gaussian prior, the output variables should therefore be Gaussian distributed. This can then be modelled by classical statistical methods such as Student's t-distribution, $\mathcal{TP}$. To achieve exchangeability, i.e. the conditioning context set is invariant to the number and order of the contexts, a naïve approach would be to simply perform sum/average operation similar to [18]. Alternatively, a Recurrent Neural Network (RNN) update scheme [11], can also be used, where the covariance matrix of the conditional image set is made to be simple (i.e. the diagonal is parameterised by μ, with upper and lower triangle parameterised by σ), each image therefore has an identical mean and variance to one other.

During testing, the number of context image-pose pairs is not required to be the same as the number as used during training, due to the property of exchangeability. Intuitively speaking, as more context images are supplied, the predicted image should look more similar to the target image. Contexts are first passed through the RNN to set up the distribution at the condition by updating the mean and variance, from which the samples can then be drawn. Each sample, drawn from this distribution, is then passed though the inverse Conditional RealNVP, whilst being conditioned on the query pose.

To create a dense 3D volume from very sparsely sampled 2D slices, a trained BRUNO model is queried with a dense sweep of all possible pose positions within the same Field-of-View (FoV) as training. The sparse sampled 2D slices, along with the corresponding pose, are therefore supplied as contexts for the model. Multiple samples can then be drawn from the trained distribution as possible hallucinations of the missing slice. Alternatively, it is also possible to take the mean image (i.e. average of infinite samples). With no contexts supplied, samples are drawn from the prior distribution. This can be used to create organ and/or volume atlases and for manual model validation, as the trained distribution is an average of all training volumes. Patient specific missing slices and extrapolated anatomy can be generated if context images are supplied. Samples are then drawn from the posterior as the distribution is conditioned on the contexts.

3   Experiments and Results

To validate the trained model, high resolution 3D volumes from the test set are decomposed into individual 2D slices with their corresponding one-hot pose vector. A sparse set of slices (between 1 to 10) are used as contexts. The model is then queried with the pose vectors, the generated slices are then compared to the target image using Cross-Correlation (CC) and Structural Similarity Index (SSIM). CC measures the similarity of pixel-wise intensities, whereas SSIM uses a combination luminance, contrast and structure to assess the image quality.

The first set of experiments used brain MRI and thorax CT images. 85 healthy brains were selected from the Alzheimer's Disease Neuroimaging Initiative (ADNI) database. These were split 70 for training and 15 for validation, with $K = 120$ axial slices extracted from the middle 75% portion of the brain. The CT images were split 50 for training and 8 for validation, with $K = 100$ axial slices extracted from the middle 50% portion of the scan volume. Both data-sets are isotropic with spacing of $1\,mm \times 1\,mm \times 1\,mm$. Each 2D slice, x, are of size 218×218 for MRI brain and 200×200 for CT thorax, and are further down-sampled to 64×64. An additional isotropic fetal brain MRI data-set, with spacing of $1\,mm \times 1\,mm \times 1\,mm$, was used for Initial Experiments and Exp2. 270 brains ranging from 40 to 43 Gestational Age (GA) were selected; split 250 for training and 20 for testing. $K = 80$ axial slices were then extracted from the middle 65% of the brain, each 2D slice of size 160×160 is further down-sampled to 64×64 for training and inference.

The pose is formulated as one-hot vector of length K, to represent the slice number of the scan stack. For all experiments, $M = 9$; a sequence of eight image-pose pairs are used as context, with the 1 remaining as query pose and target image. Contexts are randomly sampled from the entire stack during training phase. However, during testing, the contexts are strategically selected so that they cover an approximate even distribution across the scan stack. For each query pose, 100 samples were drawn from the posterior distribution and compared to the target image using SSIM and Cross-Correlation. An average is then taken across all slices for a volume average, and further averaged across all test subjects.

3.1   Initial Experiments

To compare with existing baseline models, GQN and C-VAE architectures were used to build mental maps using the Fetal brain data-set. Both GQN and C-VAE models have been trained with 4 context image-pose pairs.

As official code for the Generative Query Network have not been published, a reputable public reimplementation [8] was used instead. The implementation has been validated to be correctly functioning, as it has been successfully tested on several, but simple, official GQN data-sets. Performance however was not able to match the results published by DeepMind, as the architecture for the DRAW-LSTM was not disclosed. Only the baseline architecture was used.

A naïve Conditional Variational AutoEncoder was also implemented as a baseline. The architecture follows the structure of a standard U-Net [15], but without skip connections, and with 4 scaled resolutions and 2 convolutional layers at each resolution. It has also been formulated such that the input is a pose vector, the condition is a set of encoded context images, and the output is the generated image. Contexts are first passed through a tower encoder network, same as the GQN, to encode each image-pose pair in latent representation. All Contexts are then averaged together in latent space to maintain order invariance. During inference, the latent vector z is sampled from a unit Gaussian distribution to feed the generator whilst being conditioned on contexts as well as queried pose.

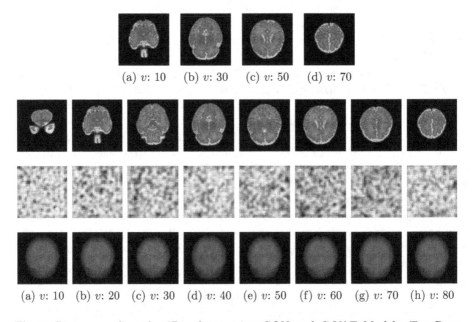

(a) v: 10   (b) v: 30   (c) v: 50   (d) v: 70

(a) v: 10   (b) v: 20   (c) v: 30   (d) v: 40   (e) v: 50   (f) v: 60   (g) v: 70   (h) v: 80

Fig. 2. Dense sampling the 3D volume using GQN and C-VAE Models. Top Row: Context slices from one particular subject at pose position v. Second Row: Ground Truth Images. Third Row: GQN predicted slice at pose position v. Bottom Row: C-VAE predicted slice at pose position v.

Both models were not able to achieve satisifiable results, as seen in Fig. 2. Due to the complex nature of the brain structure, the GQN model was not able to generalise, and predicted static for all slices. The generated images by the C-VAE model seem to resemble an average of all input context slices with variational noise on top. The experiments also shown that the models do not easily converge, as the latent distribution is far from the prior distribution as measured by the KL divergence. This is notably evident with the GQN, where the KL divergence in the generator module is often very high. Images generated by either method are corrupt.

3.2   Exp1: ADNI MRI and Thorax CT

The first set of experiments were used to evaluate the performance of the BRUNO architecture. Figure 3a and b shows the average SSIM and Cross-Correlation of the generated dense sampled volume compared to the original high resolution volume across all test subjects. 4 experiments were ran with increasing number of contexts; 1, 3, 5 and 9. The contexts selected are sparsely spread to maximise the coverage across the dense volume. In Fig. 3a and b it can be seen that as the number of contexts increases, the reconstructed volume becomes closer to the ground truth volume in similarity.

	Context Images					Context Images			
	1	3	5	9		1	3	5	9
SSIM	0.695	0.719	0.733	0.736	**SSIM**	0.779	0.786	0.787	0.791
CC	0.913	0.917	0.919	0.922	**CC**	0.802	0.815	0.832	0.833
(a) ADNI MRI Data-set					(b) Thorax CT Data-set				

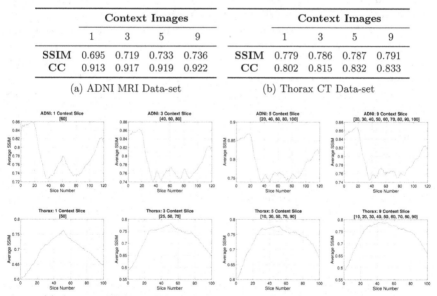

(c) Average SSIM vs Slice Number w.r.t. Number of Context Images
Top Row: ADNI MRI Data-set, Bottom Row: Thorax CT Data-set

Fig. 3. Table and figures for the results of experiment 1

Figure 3c shows the average SSIM of each slice across all test subjects for ADNI and Thorax data-sets. A distinct peak in SSIM is perceivable where a query slice aligns with a supplied context slice. This confirms the notion that conditional contexts correctly steer the posterior to a particular part of the distribution. High SSIM for ADNI 0 to 20 and 100 to 120 are the edge cases, where a majority of the content are background. The inverse is the case for thorax, where slices approaching top and bottom have high variability in structure, thus reducing the SSIM.

3.3   Exp2: Fetal Brain Template Volume

The second set of experiments has evaluated the usefulness of the proposed approach for fetal MRI reconstruction: State-of-the-art iterative image-based reconstruction methods, e.g. Slice-to-Volume Reconstruction (SVR) [9], often require a good initialisation volume for the initial target registration. In clinical setting, especially in fetal MRI, volumes are often motion corrupted if the fetus is awake and constantly moving during image acquisition. Neighbouring slices of the volumes are therefore incoherent and in disarray. In this experiment, BRUNO is used to create the initial registration target volume for 2D to 3D fetal brain reconstruction. During fetal MRI, a few images are often acquired in parallel (usually four, spatially far apart images, at once). These image batches can be used as conditional contexts for BRUNO. Due to fast parallel acquisition, the slices can be assumed to be aligned and motion free.

As with the first set of experiments, the performance of BRUNO is evaluated with varying context slices. In total, there are 80 slices in the dense fetal brain volume. Figure 4a below shows the number of contexts, with the corresponding slice numbers, that is used during inference. Figure 4b shows the SSIM and CC of average reconstructed SVR initialisation volumes. Like as in the first

Number of Contexts	Slice Number, k ($K = 80$)
1	[40]
3	[20,40,60]
4	[10,30,50,70]
7	[10,20,30,40,50,60.70]

(a) Selected Slices as Context

	Context Images			
	1	3	4	7
SSIM	0.665	0.676	0.679	0.684
CC	0.868	0.875	0.876	0.880

(b) SSIM/CC of Reconstructed Volume

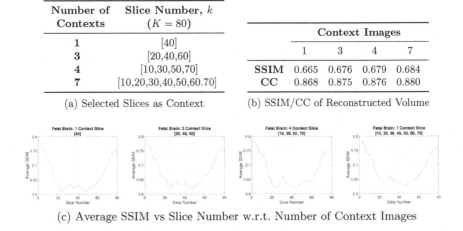

(c) Average SSIM vs Slice Number w.r.t. Number of Context Images

Fig. 4. Table and figures for the results of experiment 2

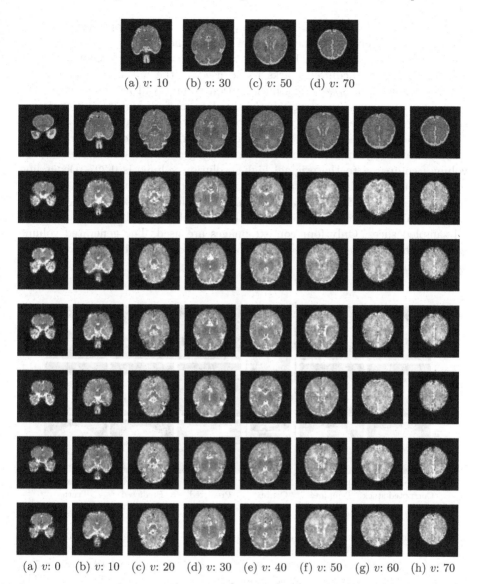

(a) v: 10   (b) v: 30   (c) v: 50   (d) v: 70

(a) v: 0   (b) v: 10   (c) v: 20   (d) v: 30   (e) v: 40   (f) v: 50   (g) v: 60   (h) v: 70

Fig. 5. A Conditional BRUNO architecture, consisting of a trainable bijector and Student's t-process statistical modelling. Top Row: Context slices from one particular test subject at pose position v, Second Row: Ground Truth Images, Third Row On-wards: six predicted slices at pose position v sampled from the posterior distribution conditioned by the contexts.

experiments, as the number of context images increase, the average SSIM and Cross-Correlation increases. Distinct peaks in SSIM are also present where a query slice approaches a supplied context slice.

Figure 5 shows more examples of samples drawn from the posterior distribution that have been conditioned by the 4 context slices.

Fetuses can move up to 20 mm [9] when active. For SVR, a Gaussian average is taken of all acquired stacks to be used as the initial registration target. If motion is too severe this initialisation volume will not be sufficient for subsequent iterative reconstruction. To simulate motion, high resolution reconstructed volumes are synthetically motion corrupted to be used as a base line. Three orthogonal stacks are made with 10 mm of random motion in translation only. On average, the SSIM of Gaussian averaged volumes motion corrupted volumes compared to the original high resolution volume is 0.297. Depending on the robustness of the SVR algorithm used, reconstruction may, or may not, be possible. A BRUNO generated volume is made by densely and repeatedly sampling all possible pose positions, with the final volume being an average of all sampled slices. Only four context images are used. The generated volume was able to achieve an SSIM of 0.679 compared to the original high resolution volume, this is shown in Fig. 6.

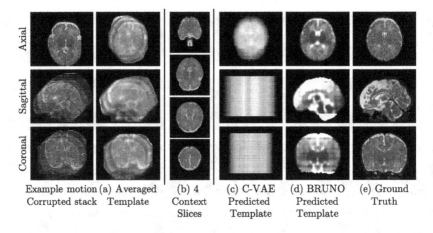

Example motion (a) Averaged   (b) 4   (c) C-VAE   (d) BRUNO   (e) Ground
Corrupted stack   Template   Context   Predicted   Predicted   Truth
   Slices   Template   Template

Fig. 6. For motion compensation of fetal and neonatal MRI, averaged template volumes (a) from all acquired slices (several 100) are used to initialise SVR [9]. Only four images were used (b) from an image batch that has been acquired in parallel, (common in fetal and neonatal MRI, i.e. without motion corruption between the four images). A C-VAE is not able to predict a template volume from these four images (c), while our approach using BRUNO predicts a reasonable volume (d) compared to the ground truth (e).

4   Conclusion and Discussion

This paper introduces the idea of using Deep Neural Networks and Bayesian Deep Learning to build mental maps of anatomy of various medical volumes. A conditional generative model, based on the BRUNO architecture, is trained on existing high resolution 3D volumes. It can be used to create patient specific volumes by densely querying all possible pose positions, whilst conditioned by a few existing slices that are used as contexts. *Exp1* shows that BRUNO is able to reconstruct MRI brain volumes with an SSIM of 0.7 and Thorax volumes with an SSIM of 0.8 compared to the original high resolution ground truth. *Exp2* shows a specific use case for BRUNO to generate initial target volumes for 2D to 3D fetal brain MRI reconstruction.

Future work will be to investigate further into the framework, and improve image quality generation as well as to introduce more Degrees-of-Freedom (DoF). In the current implementation, BRUNO is able to successfully traverse single DoF, and is applicable for use cases such CT and MRI. Increased DoF, with added rotations and translations, can be particularly valuable for modalities such as freehand ultrasound, with applications for Reinforcement Learning and 3D scene exploration.

Acknowledgements. We thank The Wellcome Trust IEH Award iFind project [102431], Innovate UK: London Medical Imaging & Artificial Intelligence Centre for Value-Based Healthcare [104691], and NVIDIA for their GPU donations. The data used in the preparation of this article were obtained from the Alzheimer's Disease Neuroimaging Initiative (ADNI) database (http://adni.loni.usc.edu). Fetal brain data were accessed only with informed consent, subject to approval and formal Data Sharing Agreement. We also like to thank Ira, author of BRUNO and Conditional BRUNO, for the valuable discussions.

References

1. Cerrolaza, J.J., et al.: 3D fetal skull reconstruction from 2DUS via deep conditional generative networks. In: Frangi, A.F., Schnabel, J.A., Davatzikos, C., Alberola-López, C., Fichtinger, G. (eds.) MICCAI 2018. LNCS, vol. 11070, pp. 383–391. Springer, Cham (2018). https://doi.org/10.1007/978-3-030-00928-1_44
2. Choy, C.B., Xu, D., Gwak, J.Y., Chen, K., Savarese, S.: 3D-R2N2: a unified approach for single and multi-view 3D object reconstruction. In: Leibe, B., Matas, J., Sebe, N., Welling, M. (eds.) ECCV 2016. LNCS, vol. 9912, pp. 628–644. Springer, Cham (2016). https://doi.org/10.1007/978-3-319-46484-8_38
3. Ding, F., Leow, W.K., Wang, S.-C.: Segmentation of 3D CT volume images using a single 2D atlas. In: Liu, Y., Jiang, T., Zhang, C. (eds.) CVBIA 2005. LNCS, vol. 3765, pp. 459–468. Springer, Heidelberg (2005). https://doi.org/10.1007/11569541_46
4. Dinh, L., et al.: Density estimation using real NVP. CoRR abs/1605.08803 (2016)
5. Ehlke, M., et al.: Fast generation of virtual X-ray images for reconstruction of 3D anatomy. IEEE Trans. Vis. Comput. Graph. **19**(12), 2673–2682 (2013)
6. Eslami, S.M.A., et al.: Neural scene representation and rendering. Science **360**(6394), 1204–1210 (2018). https://science.sciencemag.org/content/360/6394/1204
7. Goodfellow, I., et al.: Generative adversarial nets. In: NIPS. pp. 2672–2680 (2014)
8. Groth, O.: ogroth/tf-gqn, June 2019. https://github.com/ogroth/tf-gqn
9. Kainz, B., et al.: Fast volume reconstruction from motion corrupted stacks of 2D slices. IEEE Trans. Med. Imaging **34**(9), 1901–1913 (2015)
10. Kingma, D.P., et al.: Auto-encoding variational bayes. CoRR abs/1312.6114 (2013)

11. Korshunova, I., et al.: Conditional BRUNO: a deep recurrent process for exchangeable labelled data. In: Bayesian Deep Learning NeurIPS Workshop (2018)
12. Kunter, M., et al.: Unsupervised object segmentation for 2D to 3D conversion. In: Stereoscopic Displays and Applications XX, vol. 7237, p. 72371B (2009)
13. Papamakarios, G., Pavlakou, T., Murray, I.: Masked autoregressive flow for density estimation. In: NIPS, pp. 2338–2347 (2017)
14. Rezende, D.J., et al.: Variational inference with normalizing flows. In: ICML. JMLR Workshop and Conference Proceedings, vol. 37, pp. 1530–1538. JMLR.org (2015)
15. Ronneberger, O., Fischer, P., Brox, T.: U-Net: convolutional networks for biomedical image segmentation. In: Navab, N., Hornegger, J., Wells, W.M., Frangi, A.F. (eds.) MICCAI 2015. LNCS, vol. 9351, pp. 234–241. Springer, Cham (2015). https://doi.org/10.1007/978-3-319-24574-4_28
16. Sohn, K., et al.: Learning structured output representation using deep conditional generative models. In: NeurIPS, pp. 3483–3491 (2015)
17. Wang, L., et al.: Unsupervised 3D reconstruction from a single image via adversarial learning. CoRR abs/1711.09312 (2017)
18. Zaheer, M., et al.: Deep Sets. In: NeurIPS, pp. 3394–3404 (2017)

Spatiotemporal PET Reconstruction Using ML-EM with Learned Diffeomorphic Deformation

Ozan Öktem[1], Camille Pouchol[1(✉)], and Olivier Verdier[1,2]

[1] Department of Mathematics, KTH Royal Institute of Technology,
100 44 Stockholm, Sweden
ozan@kth.se

[2] Department of Computing, Mathematics and Physics, Western Norway
University of Applied Sciences, Bergen, Norway

Abstract. Patient movement in emission tomography deteriorates reconstruction quality because of motion blur. Gating the data improves the situation somewhat: each gate contains a movement phase which is approximately stationary. A standard method is to use only the data from a few gates, with little movement between them. However, the corresponding loss of data entails an increase of noise. Motion correction algorithms have been implemented to take into account all the gated data, but they do not scale well in computation time, especially not in 3D. We propose a novel motion correction algorithm which addresses the scalability issue. Our approach is to combine an enhanced ML-EM algorithm with deep learning based movement registration. The training is unsupervised, and with artificial data. We expect this approach to scale very well to higher resolutions and to 3D, as the overall cost of our algorithm is only marginally greater than that of a standard ML-EM algorithm. We show that we can significantly decrease the noise corresponding to a limited number of gates.

Keywords: Emission tomography · Motion correction · Deep learning

1 Introduction

Positron emission tomography (PET) is a molecular imaging technology where a radioactive tracer is administered to a patient. The tracer is an x-ray source that emits pairs of photons travelling into opposite directions, and the PET scanner is an arrangement of detectors for detecting such photon pairs (coincidence events). The goal is then to recover the spatial distribution of the tracer (activity map) from these coincidence events.

Acquiring a sufficient amount of coincidence events takes time, typically twenty to forty minutes depending on the detector efficiency and the size of the region being imaged. Organs, such as the heart and lungs, move during the PET data acquisition, so the activity map one seeks to recover in PET imaging is a spatiotemporal quantity. Failure to account for the temporal variability during reconstruction results in a deteriorated PET image.

© Springer Nature Switzerland AG 2019
F. Knoll et al. (Eds.): MLMIR 2019, LNCS 11905, pp. 151–162, 2019.
https://doi.org/10.1007/978-3-030-33843-5_14

1.1   Survey of Existing Works

Most approaches that consider motion in PET image reconstruction assume access to gated PET data. Here, PET data is subdivided into subsets where the coincidence data is from the activity map in a specific temporal state. For cardiac and respiratory motion, gated data would correspond to decomposing the entire dataset into parts that represent different breathing and/or cardiac phases. Hence, the activity associated to each gate can be assumed to be stationary, but data in the gates also suffer from a relatively low signal-to-noise ratio since they only contain a small portion of the coincidence events.

A straightforward approach based on gated data is to recover each temporal state of the activity independently of each other (frame-by-frame reconstruction). This does not account for the temporal dynamics of the activity, instead it reduces the spatiotemporal reconstruction problem into a sequence of independent stationary reconstruction problems, which in PET is done by ML-EM [21] (or a variant thereof, like OSEM [14]). Spatiotemporal reconstruction refers to methods that instead take the temporal dynamics into account. Several approaches have been proposed where most rely on estimating a motion model prior to reconstruction, see [8,10,18,19] for survey.

In this paper, the proposed method falls into the family of algorithms that, contrarily to those based on a priori built motion models, jointly estimate the image and motion, directly from the full set of measured data. An objective function is optimised with respect to two arguments: image and motion. Hence, only one image with the full statistic is reconstructed. Given the close relationship between the image reconstruction and motion estimation steps, a simultaneous method of estimating the two is better able to reduce motion blur and compensate for poor signal-to-noise ratios and to improve the accuracy of the estimated motion [11,12].

In the latter works, one performs a two-step minimisation of a joint energy functional term (which includes both image likelihood and motion-matching terms). The method chosen by Jacobson and Fessler [15,16], referred to as *joint estimation with deformation modelling*, is based on maximising the likelihood for a parametric Poisson model for gated PET measurements. Motion (from gate to gate) is defined by a set of deformation parameters. A similar motion-aware likelihood function was used by Blume and colleagues [5], although using a distinct optimisation scheme and depicting more convincing results. In this context one may also consult [23], which compares three approaches for joint reconstruction of image and motion.

An alternative is to consider motion models derived from deformations modelled by diffeomorphisms, as obtained from example through the LDDMM framework [22]. Here, one can calculate regularising functionals that incorporate such deformations. Finally, [9] provides an overview of variational shape models as applied to the registration and segmentation problems. These could also be coupled with variational regularisation methods for image reconstruction.

The main drawback of all these methods, however, is the relatively high computational costs involved in the joint reconstruction approach.

1.2   Proposed Method

In this paper, we develop a joint reconstruction method based on the minimisation of a suitable functional. The main novelty of our work is the scalability of the resulting algorithm, as its complexity is of the order of the usual ML-EM algorithm. Images are indeed estimated using a generalised ML-EM algorithm. Motion estimation, with deformations modelled by diffeomorphisms, is based on the unsupervised deep learning framework voxelmorph [6]. That is, we make use of a pre-trained neural network which performs direct image registration, i.e., the network finds a diffeomorphism which, given two images, deforms the first one into the second.

Interestingly, one single outer iteration of our algorithm is close to the recently proposed approach [17]. Thus, it generalises the previous work and shows that it can be interpreted in the framework of an optimisation problem.

The results of the proposed method are tested on the Derenzo phantom, and shown to recover a significant part of the information lost when one uses gate-by-gate reconstruction.

2   Methods

2.1   Mathematical Background

ML-EM Algorithm [21]. Let us consider the statistical model

$$g \sim \text{Poisson}(Af),$$

where f is the unknown image, and g is the acquired data—a vector of $\mathbb{R}^d$; this models the physics of *stationary* PET with forward operator A.

The ML-EM algorithm solves the corresponding *maximum likelihood problem*, which amounts to minimising the divergence $d_{KL}(g||Af)$, defined for two non-negative vectors u, v in $\mathbb{R}^d$ by

$$d_{KL}(u||v) := \sum_{j=1}^{d} \left(v_j - u_j - u_j \log \left(\frac{u_j}{v_j} \right) \right).$$

The ML-EM algorithm is an iterative solver with update

$$f^{(n+1)} := \frac{f^{(n)}}{A^T 1} A^T \left(\frac{g}{Af^{(n)}} \right), \tag{1}$$

starting from an initial guess $f^{(0)}$, usually $f^{(0)} = 1$.

Diffeomorphisms Acting on Images. Viewing images as elements of $X := L^2(\Omega)$, i.e., square-integrable functions on a compact $\Omega \subset \mathbb{R}^p$ with $p = 2$ or $p = 3$, we model motion as an appropriate *group action* of diffeomorphisms onto images. In this paper, given a diffeomorphism $\psi\colon \Omega \to \Omega$, we will use the specific definition $\mathcal{W}_\psi\colon X \mapsto X$ as the *intensity-preserving* action

$$\mathcal{W}_\psi f(x) := f(\psi^{-1}(x)).$$

Note that our approach is, however, general, and we could have used the *mass-preserving* action instead, namely

$$\widetilde{\mathcal{W}}_\psi f(x) := |D\psi^{-1}(x)| f(\psi^{-1}(x)). \tag{2}$$

We will parameterise diffeomorphisms by exponentials of (stationary) vector fields, i.e., $\psi = \exp(v)$, where the exponential $\exp(v)$ of a vector field v is defined as $\psi(1, \cdot)$, where $\psi(t, \cdot)$ solves the differential equation $\frac{\partial \psi}{\partial t}(t, \cdot) = v(\psi(t, \cdot))$, with initial condition $\psi(0, \cdot) = \mathrm{Id}$.

Image Registration. The (direct) image registration problem consists in deforming a *template* f_1 into a *target* f_2, i.e., finding a diffeomorphism ψ such that $\mathcal{W}_\psi f_1 \approx f_2$. This is usually done by minimising a functional of the form

$$\operatorname*{arg\,min}_{\psi}\ d_2(f_2, \mathcal{W}_\psi f_1) + \lambda \mathcal{R}(\psi), \tag{3}$$

where d_2 is the L^2-distance on X, $\mathcal{R}$ is a regularisation term on diffeomorphisms that is discussed in Subsect. 2.3, and λ is a regularisation parameter.

2.2   General Approach

Modelling. We are given *gated* data in $N + 1$ different gates, corrupted by Poisson noise. For g_i denoting the data, f_i the images in each gate and A the forward operator, we thus assume

$$g_i \sim \mathrm{Poisson}(Af_i),\ i = 0, \ldots, N.$$

We also assume that for $i = 1, \ldots, N$, two consecutive images f_{i-1} and f_i are related by the statistical model

$$f_i = \mathcal{W}_{\psi_i} f_{i-1} + e_i,$$

where $\psi_i\colon \Omega \to \Omega$ is the exponential of a vector field following a given probability law (see (8)) and e_i is a X-valued random variable.

Variational Problem. We now define the variational problem associated to the inverse problem of finding both the images f_i and diffeomorphisms ψ_i from the data g_i. It reads

$$\operatorname*{arg\,min}_{(f_i),(\psi_i)}\ J(f_0, \ldots, f_N, \psi_1, \ldots, \psi_N), \tag{4}$$

where

$$J(f_i, \psi_i) := \sum_{i=0}^{N} d_{KL}(g_i \| Af_i) + \sum_{i=1}^{N} \Big(d_2(f_i, \mathcal{W}_{\psi_i} f_{i-1}) + \lambda \mathcal{R}(\psi_i) \Big).$$

General Algorithm. We solve the variational problem (4) by an *intertwined* method, which consists in alternating between estimating the diffeomorphisms (the *motion estimation step*), and the images f_i (the *reconstruction step*).

The images are first initialised by solving the maximum likelihood problem $\arg\min_{f_i}(d_{KL}(g_i\|Af_i))$, associated to $g_i = \text{Poisson}(Af_i)$ in each gate. This is done by the algorithm ML-EM (1), yielding estimates f_i^0, $i = 0, \ldots, N$.

For a given estimate of images f_i^k, the motion estimation part consists in solving

$$\arg\min_{(\psi_i)} \sum_{i=1}^{N} \left(d_2(f_i^k, \mathcal{W}_{\psi_i} f_{i-1}^k) + \lambda \mathcal{R}(\psi_i)\right),$$

which in turn can be decomposed into N problems of the form

$$\arg\min_{\psi_i}\ d_2(f_i^k, \mathcal{W}_{\psi_i} f_{i-1}^k) + \lambda \mathcal{R}(\psi_i),\ i = 1, \ldots, N. \tag{5}$$

Note that each of these becomes an *image registration problem*, as we are looking for a diffeomorphism ψ_i^{k+1} matching the template f_{i-1}^k against the target f_i^k.

For the reconstruction part, we assume $f_i^k \approx \mathcal{W}_{\psi_i^{k+1}} f_{i-1}^k$ for $i = 1, \ldots, N$ and neglect the N corresponding d_2 terms. The minimisation problem thus becomes

$$\arg\min_{(f_i)} \sum_{i=0}^{N} d_{KL}(g_i\|Af_i).$$

We then focus on a particular gate, say the zero'th gate, and use $f_i^k \approx \mathcal{W}_{\psi_i^{k+1}} f_i^k$ to obtain the optimisation problem with f_0 as the only variable:

$$\arg\min_{f_0} \sum_{i=0}^{N} d_{KL}(g_i\|A_i f_0). \tag{6}$$

where

$$A_i := A\mathcal{W}_{\phi_i} \tag{7}$$

and we have used the notation $\phi_i := \psi_i \circ \cdots \circ \psi_1$ for $i = 1, \ldots, N$. Solving the above yields a next estimate f_0^{k+1} for f_0. All the images f_i^{k+1} are then obtained by $f_i^{k+1} = \mathcal{W}_{\psi_i^{k+1}} f_{i-1}^{k+1}$, $i = 1, \ldots, N$.

It now only remains to explain how the optimisation problems (5) and (6) are solved, which is the topic of the next subsections.

2.3 Motion Estimation

The motion estimation problem (5), can be rewritten for two generic images f_1 and f_2 as

$$\arg\min_{v}\ d_2(f_2, \mathcal{W}_{\exp(v)} f_1) + \lambda \mathcal{R}(v), \tag{8}$$

where we parameterise the diffeomorphisms by exponentials of stationary vector fields v.

To solve this direct image registration problem, we use the `voxelmorph` unsupervised deep learning approach [6], where a neural network parameterises a function $(f_1, f_2) \mapsto v$. That neural network is itself based on the network architecture `Unet` [20]. We keep the architecture of `voxelmorph`, with the same hyperparameters and specific regularisation functional $\mathcal{R}$ given in [6]. Once trained, the network produces a mapping matching any two images f_1, f_2, which we denote

$$\gamma(f_1, f_2) := \exp(v(f_1, f_2)). \tag{9}$$

Training. In [6], the network is trained on tuples of images (f_1, f_2) coming from brain MRI scans. We use instead *synthetic* data: tuples of images (f_1, f_2) generated on the fly.

We generate training images as follows. A random image f_1 consists of a Poisson random number of ellipsoids [3,4]. The centre of each ellipsoid has uniform distribution inside the central part of the domain Ω, the principal axes have exponential distribution, and the orientation follows a uniform distribution. We apply a mask vanishing at the boundary to avoid boundary effects. when diffeomorphisms are applied. We generate random vector fields v using a Gaussian random field with radial basis function kernel, with appropriate scale and typical size. The training image f_2 is then $f_2 = \mathcal{W}_{\exp(v)} f_1$. We show in Fig. 1 a sample of images f_1, f_2 and vector field v generated as above.

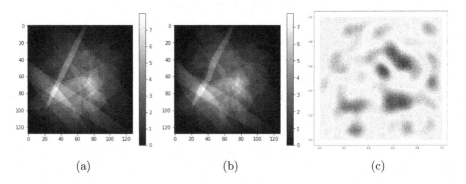

(a)   (b)   (c)

Fig. 1. Example of a 2D synthetic tuple of images f_1 (a) and f_2 (b), related by $f_2 = \mathcal{W}_{\exp(v)} f_1$ for the intensity-preserving action, with v plotted in (c).

2.4   Reconstruction

We now focus on the reconstruction problem (6) which we solve using a reformulation of ML-EM. Given operators A_i, we can simply write ML-EM for the compound operator $A = (A_0, \ldots, A_N)$ which yields

$$f_0^{(n+1)} = \frac{f_0^{(n)}}{\sum_{i=0}^{N} A_i^T 1} \sum_{i=0}^{N} A_i^T \left(\frac{g_i}{A_i f_0^{(n)}} \right), \tag{10}$$

for an initial guess $f_0^{(0)}$. We call this algorithm "M-ML-EM" to avoid the confusion with the vanilla ML-EM algorithm (1). We use this algorithm with A_i defined in (7). Note that this algorithm has been used in [13] for the particular case of the intensity-preserving action. The computation of A_i^T requires the computation of $\mathcal{W}_{\phi_i}^T$. We achieve this by using the identity $W_\phi^T = \widetilde{\mathcal{W}}_{\phi^{-1}}$ valid for any diffeomorphism ϕ, where $\widetilde{\mathcal{W}}$ denotes the mass-preserving action (2).

2.5 Full Algorithm

We summarise the algorithm with all the necessary details in Algorithm 1.

Algorithm 1. Full Algorithm

Choose the outer number of iterates n_{outer}, the inner number of iterates n_{inner} for M-ML-EM, and n_{init}, the number of iterates for vanilla ML-EM.

 for $i \leftarrow 0, \dots, N$ **do**
  $f_i \leftarrow \text{ML-EM}(A, g_i, n_{\text{init}})$ ▷ Iterates of (1)
 end for
 for $k \leftarrow 1, \dots, n_{\text{outer}}$ **do**
  for $i \leftarrow 1, \dots, N$ **do**
   $\psi_i \leftarrow \gamma(f_{i-1}, f_i)$ ▷ Network registration (9)
  end for
  $W_0 \leftarrow \text{Id}$
  for $i \leftarrow 1, \dots, N$ **do**
   $W_i \leftarrow W_{\psi_i} W_{i-1}$
   $A_i \leftarrow A W_i$
  end for
  $A_0 \leftarrow A$
  $f_0 = \text{M-ML-EM}(\{A_j\}_{j=0,\dots,N}, \{g_j\}_{j=0,\dots,N}, n_{\text{inner}})$ ▷ Iterates of (10)
  for $i \leftarrow 1, \dots, N$ **do**
   $f_i \leftarrow W_i f_0$
  end for
 end for
The outcome is f_0.

2.6 Complexity

Evaluating vector fields with the network is negligible when compared to ML-EM or M-ML-EM iterations. Each iteration is itself controlled by the time t required to compute an expression of the form $A^T(\frac{g}{Af})$. Since M-ML-EM sums these quantities N times, an iteration of it is of the order of $N \times t$. Note that evaluating the denominator in (10) (which involves sums of $A^T 1$) does not take more time than evaluating the denominator in ML-EM since $A^T 1$ can be computed off-line.

3   Results

3.1   Derenzo Phantom

We present experiments with the Derenzo phantom, with image size 192×192. Although this phantom is made of ellipses, we stress that they are very different from the data used to train the network, compare Figs. 1 and 2a.

This phantom is then deformed successively with the intensity preserving action by exponentials of vector fields, where each vector field is drawn from the same distribution used to train the network. For the experiments, we use $N = 3$, which amounts to four gates, and we want to recover the image in the initial gate. The resulting four phantoms are presented in Fig. 2.

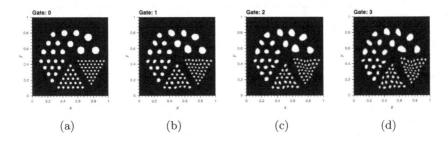

| (a) | (b) | (c) | (d) |

Fig. 2. Derenzo phantom in four different gates.

The forward operator A is a 2D PET operator with 108 angles (views) and 250 tangential positions. The noisy data is $\text{Poisson}(A(tf))$ for each image f, where t is the acquisition time and thus controls the noise level.

For the phantoms in Fig. 2, we choose $t = 60$. This noise level gives rise to typical optimal numbers of iterates for ML-EM which are of the same order of magnitude as the ones in clinic applications. Note that all images are multiplied by the same time factor, which amounts to assuming that acquisition time is roughly the same in each gate.

3.2   Methods Without Motion Correction

We compare our method with two simple reconstruction methods (simple because without motion correction) for images with gated data:

- Either one aggregates the whole data and reconstructs from ML-EM, leading to blurry results because of the movement.
- Or one tries and limit blur by focusing on one gate (say the first) and reconstructing only from that. Since there is less data, the result is noisier.

In order to quantitatively compare these strategies, we use ML-EM for the data obtained from taking gate zero only, aggregating gates zero and one, and so on up until aggregating all the four gates. Finally, we can also estimate the best reconstruction one could hope for, that is, if there were no movement. This amounts to acquiring the phantom in the 0th gate four times longer.

The results are given in Fig. 3, where the PSNR between the estimated image and the real image in gate zero is computed at each iteration.

The results show that aggregating the gates progressively induces a drop in image quality, as measured by the PSNR. Compared to gate zero acquired four times longer, the best possible achievable gain is about 2.2 dB.

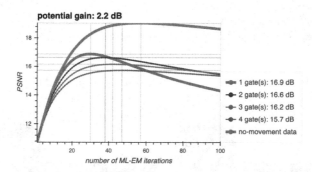

Fig. 3. PSNR for different ML-EM strategies without motion correction, and comparison with "no-movement" data, reconstructing from the initial gate acquired four times longer.

3.3 Proposed Method

We apply Algorithm 1 to the data above. It turns out that a single outer iteration is responsible for most of the improvement, so we focus on that case for presenting experiments. In other words:

1. we initialise by running some ML-EM iterations in each gate,
2. we then match the resulting images to estimate the diffeomorphisms,
3. we finally run some M-ML-EM iterations.

We plot the PSNR between the image reconstructed (in the initial gate zero) and the real image, for a given number em_iter of ML-EM iteration followed by a given number diff_iter of M-ML-EM iterations. These results are presented in Fig. 4.

We find that the optimal strategy is to iterate only a few times (six iterations in this specific experiment) with ML-EM before estimating the diffeomorphisms through M-ML-EM (42 iterations in this specific experiment). Note that this yields a total of 48 iterations which is higher than the 29 ML-EM iterations which would be optimal for reconstructing from the gate zero.

The gain in PSNR is 1.0 dB, which makes up for about 46% of the maximal gain of 2.2 dB. Reconstructions obtained from the optimal uncorrected ($n = 29$

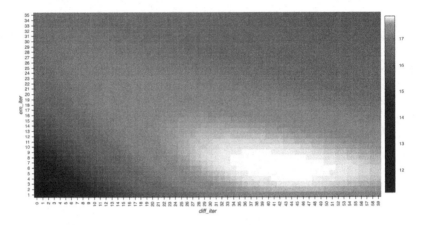

Fig. 4. PSNR for various choices of number `em_iter` of initial ML-EM iterates and number `diff_iter` of M-ML-EM iterates.

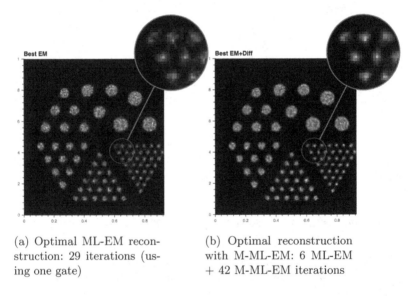

(a) Optimal ML-EM reconstruction: 29 iterations (using one gate)

(b) Optimal reconstruction with M-ML-EM: 6 ML-EM + 42 M-ML-EM iterations

Fig. 5. Optimal reconstructions of the gate zero (measured in PSNR).

iterations of ML-EM are used on gate zero) and the proposed method with the optimal number of iterations of ML-EM and M-ML-EM are presented in Fig. 5. The proposed method seems to give smoother results. The smaller discs towards the middle of the image are also better seen.

We also emphasise that these results (improvement in PSNR and optimal number of iterations) are extremely robust with respect to the randomness involved in the experiments, namely the vector fields drawn randomly as well as the Poisson noise.

3.4 Implementation Details

All computations are run in Python and use Operator Discretization Library (odl) for manipulating operators [2], neuron for warping utilities [7], which itself uses tensorflow [1]. The training was performed with voxelmorph [6].

4 Perspectives

This paper presents a new method for joint motion estimation and image reconstruction in PET. Its main advantage is its cost, similar to that of the usual ML-EM algorithm, making it scalable to clinical 4D data.

Our framework also allows for further modelling such as attenuation correction. In a future work, we consider testing this method with clinical data. This would require training the network on appropriate datasets. We also plan to generalise the approach to other group actions, such as the mass-preserving one, which is more physically relevant.

Acknowledgements. We acknowledge support from the Swedish Foundation of Strategic Research grant AM13-004.

References

1. Abadi, M., et al.: TensorFlow: large-scale machine learning on heterogeneous systems (2015). http://tensorflow.org/. Software available from tensorflow.org
2. Adler, J., Kohr, H., Öktem, O.: ODL-a Python framework for rapid prototyping in inverse problems. Royal Institute of Technology (2017)
3. Adler, J., Öktem, O.: Solving ill-posed inverse problems using iterative deep neural networks. Inverse Probl. **33**(12), 124007 (2017)
4. Adler, J., Öktem, O.: Learned primal-dual reconstruction. IEEE Trans. Med. Imaging **37**(6), 1322–1332 (2018)
5. Blume, M., Martinez-Moller, A., Keil, A., Navab, N., Rafecas, M.: Joint reconstruction of image and motion in gated positron emission tomography. IEEE Trans. Med. Imaging **29**(11), 1892–1906 (2010)
6. Dalca, A.V., Balakrishnan, G., Guttag, J., Sabuncu, M.R.: Unsupervised learning for fast probabilistic diffeomorphic registration. In: Frangi, A.F., Schnabel, J.A., Davatzikos, C., Alberola-López, C., Fichtinger, G. (eds.) MICCAI 2018. LNCS, vol. 11070, pp. 729–738. Springer, Cham (2018). https://doi.org/10.1007/978-3-030-00928-1_82
7. Dalca, A.V., Guttag, J., Sabuncu, M.R.: Anatomical priors in convolutional networks for unsupervised biomedical segmentation. In: Proceedings of the IEEE Conference on Computer Vision and Pattern Recognition, pp. 9290–9299 (2018)
8. Dawood, M., Jiang, X., Schäfers, K.P. (eds.): Correction Techniques in Emission Tomography. Series in Medical Physics and Biomedical Engineering. CRC Press, Boca Raton (2008)
9. Farag, A.A., Shalaby, A., Abd El Munim, H., Farag, A.: Variational shape representation for modeling, elastic registration and segmentation. In: Li, S., Tavares, J.M.R.S. (eds.) Shape Analysis in Medical Image Analysis. LNCVB, vol. 14, pp. 95–121. Springer, Cham (2014). https://doi.org/10.1007/978-3-319-03813-1_3

10. Gigengack, F., Jiang, X., Dawood, M., Schäfers, K.P.: Motion Correction in Thoracic Positron Emission Tomography. Springer, Cham (2015). https://doi.org/10.1007/978-3-319-08392-6
11. Gilland, D.R., Mair, B.A., Bowsher, J.E., Jaszczak, R.J.: Simultaneous reconstruction and motion estimation for gated cardiac ECT. IEEE Trans. Nucl. Sci. **49**(5), 2344–2349 (2002)
12. Gravier, E., Yang, Y., King, M.A., Jin, M.: Fully 4D motion-compensated reconstruction of cardiac SPECT images. Phys. Med. Biol. **51**(18), 4603–4619 (2006)
13. Hinkle, J., Szegedi, M., Wang, B., Salter, B., Joshi, S.: 4D CT image reconstruction with diffeomorphic motion model. Med. Image Anal. **16**(6), 1307–1316 (2012)
14. Hudson, H.M., Larkin, R.S.: Accelerated image reconstruction using ordered subsets of projection data. IEEE Trans. Med. Imaging **13**(4), 601–609 (1994)
15. Jacobson, M.W., Fessler, J.A.: Joint estimation of image and deformation parameters in motion-corrected PET. In: 2003 IEEE Nuclear Science Symposium Conference Record, pp. 3290–3294 (2003)
16. Jacobson, M.W., Fessler, J.A.: Joint estimation of respiratory motion and activity in 4D PET using CT side information. In: 3rd IEEE International Symposium on Biomedical Imaging: Nano to Macro, Arlington, VA, 6–9 April 2006, pp. 275–278 (2006)
17. Li, T., Zhang, M., Qi, W., Asma, E., Qi, J.: Motion correction of respiratory-gated pet image using deep learning based image registration framework. In: 15th International Meeting on Fully Three-Dimensional Image Reconstruction in Radiology and Nuclear Medicine, vol. 11072, p. 110720Q. International Society for Optics and Photonics (2019)
18. Rahmim, A., Tang, J., Zaidi, H.: Four-dimensional image reconstruction strategies in cardiac-gated and respiratory- gated PET imaging. PET Clin. **8**(1), 51–67 (2013)
19. Reader, A.J., Verhaeghe, J.: 4D image reconstruction for emission tomography. Phys. Med. Biol. **59**(22), R371–R418 (2014)
20. Ronneberger, O., Fischer, P., Brox, T.: U-Net: convolutional networks for biomedical image segmentation. In: Navab, N., Hornegger, J., Wells, W.M., Frangi, A.F. (eds.) MICCAI 2015. LNCS, vol. 9351, pp. 234–241. Springer, Cham (2015). https://doi.org/10.1007/978-3-319-24574-4_28
21. Shepp, L.A., Vardi, Y.: Maximum likelihood reconstruction for emission tomography. IEEE Trans. Med. Imaging **1**(2), 113–122 (1982)
22. Younes, L.: Shapes and Diffeomorphisms. Applied Mathematical Sciences, vol. 171. Springer, Heidelberg (2010). https://doi.org/10.1007/978-3-642-12055-8
23. Zhang, Y., Ghodrati, A., Brooks, D.H.: An analytical comparison of three spatiotemporal regularization methods for dynamic linear inverse problems in a common statistical framework. Inverse Probl. **21**(1), 357–382 (2005)

Stain Style Transfer Using Transitive Adversarial Networks

Shaojin Cai[1,2], Yuyang Xue[3], Qinquan Gao[1,2,3], Min Du[1,2], Gang Chen[4],
Hejun Zhang[4], and Tong Tong[1,2,3(✉)]

[1] College of Physics and Information Engineering,
Fuzhou University, Fuzhou, China
ttraveltong@gmail.com
[2] Fujian Key Lab of Medical Instrumentation and Pharmaceutical Technology,
Fuzhou, China
[3] Imperial Vision Technology, Fuzhou, China
[4] Department of Pathology, Fujian Provincial Cancer Hospital,
The Affiliated Hospital of Fujian Medical University, Fuzhou, China

Abstract. Digitized pathological diagnosis has been in increasing demand recently. It is well known that color information is critical to the automatic and visual analysis of pathological slides. However, the color variations due to various factors not only have negative impact on pathologist's diagnosis, but also will reduce the robustness of the algorithms. The factors that cause the color differences are not only in the process of making the slices, but also in the process of digitization. Different strategies have been proposed to alleviate the color variations. Most of such techniques rely on collecting color statistics to perform color matching across images and highly dependent on a reference template slide. Since the pathological slides between hospitals are usually unpaired, these methods do not yield good matching results. In this work, we propose a novel network that we refer to as Transitive Adversarial Networks (TAN) to transfer the color information among slides from different hospitals or centers. It is not necessary for an expert to pick a representative reference slide in the proposed TAN method. We compare the proposed method with the state-of-the-art methods quantitatively and qualitatively. Compared with the state-of-the-art methods, our method yields an improvement of 0.87 dB in terms of PSNR, demonstrating the effectiveness of the proposed TAN method in stain style transfer.

Keywords: Pathological slides · Stain transfer · Color transfer · Generative adversarial networks

1 Introduction

Staining is a general process in pathology. However, the differences in raw material, staining protocols and slide scanners between labs make the appearance

© Springer Nature Switzerland AG 2019
F. Knoll et al. (Eds.): MLMIR 2019, LNCS 11905, pp. 163–172, 2019.
https://doi.org/10.1007/978-3-030-33843-5_15

of the pathological stain suffer from large variability. These variations not only affect the diagnosis of the pathologists [9], but also can hamper the performance of CAD systems [6].

As an alternative, algorithms for automated standardization of digitized whole-slide images (WSI) have been published [1,2,4,12,14,15,17,18]. These methods can be roughly divided into three categories. **Color-matching based methods** that try to match the color spectrums between the image and reference template image. Reinhard et al. [18] matched the color-channels between the image and reference template image in the LAB color space. However, this global color mapping fails in some local regions of image, as the same transformation is applied across the whole image while ignoring the independent distributions of color in different areas of the pathological image.

In addition, **Stain separation based methods** that do the normalized operations on each staining channel independently. Macenko et al. [14] proposed the stain vectors by transforming the RGB to the Optical Density (OD) space. Khan et al. [12] assigned every pixel to the specific stain component and estimated the stain matrix. Bejnordi et al. [3] thought that these methods did not take the spatial features of the tissue into account, which might lead to improper staining. Moreover, picking a good reference image requires expert knowledge and a bad reference may hamper the performance of these methods. The third group are **Deep-learning based approaches**. These methods take advantage of the Generative adversarial networks (GANs) to transfer stain style. BenTaieb [5] designed a stain normalization net based on GANs with a discriminative image analysis model on top. However, this stain style transfer model depends on a specific model for a specific task on top. Shaban [20] proposed a method which is known as StainGAN. StainGAN is based on an Unpaired Image-to-Image Translation using Cycle-Consistent Adversarial Networks (CycleGAN). Cycle-consistency allows the images to be mapped into different color models but preserving the same tissue structure.

In this paper, we propose Transitive Adversarial Networks (TAN). TAN is also based on an Unpaired Image-to-Image Translation using Cycle-Consistent Adversarial Networks (CycleGAN) [22]. We proposed a novel generator, which can result in more accurate color transfer than other generators. TAN not only eliminates the problem of picking the reference template image but also achieve much higher quality and much faster processing speed than StainGAN, making it easier to minish the stain variants and improving the diagnosis process of the pathologists and CAD system. We have compared our method with state-of-the-art methods quantitatively and qualitatively, which demonstrates superiority of the proposed method.

2   Methodology

2.1   The Framework

Our framework is illustrated in Fig. 1. TAN is an unsupervised framework based on CycleGAN [22] in stain style transfer, which allows bidirectional transference

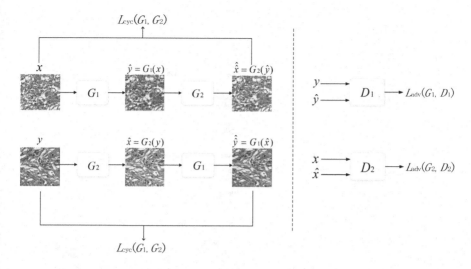

Fig. 1. The proposed framework for stain style transfer. x and y are unpaired images randomly sampled from their respective domains.

of the H&E Stain Appearance between different scanners, i.e from Aperio (A) to Hamamatsu (H) Scanner. This framework does not require paired data from different scanners. The model consists of two generators $G_1 : A \rightarrow H$ and $G_2 : H \rightarrow A$. Each generator is trained with a corresponding discriminator, D_1 and D_2. For illustration, the first pair (G_1 and D_1), try to map images from domain A to domain H. The source images x in the domain A is the input of the generator G_1, which yields generated images $\hat{y}$, $\hat{y} = G_1(x)$. Both the generated images $\hat{y}$ and the unpaired target images y in the domain H are treated as inputs of the discriminator network D_1. During the training process, G_1 and D_1 compete with each other. D_1 acts as a binary classifier, trying to distinguish the generated images $\hat{y}$ and target domain images y. Due to the adversarial training process, G_1 tries to improve the quality of the generated images $\hat{y}$ to foolish D_1. This training producer is formulated as a min-max optimization which has a adversarial loss function:

$$L_{adv}(G_1, D_1) = E_{y \sim p_{data}(y)}[log D_1(y)] + E_{x \sim p_{data}(x)}[log(1 - D_1(G_1(x)))] \quad (1)$$

Analogous to the first pair of the generator network G_1 and the discriminator network D_1, the second pair (G_2 and D_2), try to map images from the domain H to the domain A, which replaces the input images as y and the output images as x. The training producer is also formulated as a min-max optimization process, and the loss function is $L_{adv}(G_2, D_2)$:

$$L_{adv}(G_2, D_2) = E_{x \sim p_{data}(x)}[log D_2(x)] + E_{y \sim p_{data}(y)}[log(1 - D_2(G_2(y)))] \quad (2)$$

However, if the training process is merely guided by the adversarial loss, it may result in the non-convergence of the training process and lead to model

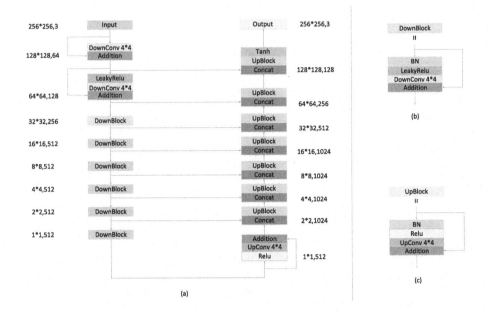

Fig. 2. The network of our proposed generator that we refer to as Trans-Net.

collapse. Several images from source domain will map to the single image in the target domain if only the adversarial loss is used. Therefore, additional training constraint on the mapping function is essential. This is achieved by adding a cycle loss, which enforces the two mapping functions, G_1 and G_2, to be cycle-consistent with each other. Generally speaking, two mapping functions should be reciprocal, for illustration, $\hat{x} = G_2(G_1(x))$, $\hat{y} = G_1(G_2(y))$. This behaviour can be achieved by adding the pixel-wise cycle-consistency loss for both generators:

$$L_{cyc}(G_1, G_2) = E_{x \sim p_{data}(x)}[\|x - G_2(G_1(x))\|_1]$$
$$+E_{y \sim p_{data}(y)}[\|y - G_1(G_2(y))\|_1] \tag{3}$$

As a result, the final loss for the whole training process can be described as:

$$L(G_1, G_2, D_1, D_2) = L_{adv}(G_1, D_1) + L_{adv}(G_2, D_2) + \lambda L_{cyc}(G_1, G_2) \tag{4}$$

2.2 Network Architectures

Generator. Compared to U-Net, we have three innovations: (i) We increase the numbers of downsampling layer and upsampling layer from 4 to 8 symmetrically, enabling the network to learn higher-level semantic information and generate more detailed context. And the experimental results demonstrate that we can produce the best images at the 8 level of downsampling. (ii) As shown in Fig. 2(b) and (c), for each downsampling layer or upsampling layer, we add the information before sampling to the sampled information innovatively. The operation can propagate the information from previous layer to next layer, which takes advantage of low-level features with low complexity and high-level features with high complexity, making it easier to get a smooth decision function with better generalization performance and produce more detailed results both in color and texture. (iii) In U-Net, convolution operations are performed twice before each sampling operation. But in Trans-Net, we delete the two convolution operations before each sampling layer. There is no convolution layer between adjacent sampling layers, avoiding the overfitting problem. And the network only includes 16 convolution layers which is 7 layers less than U-Net, making it present a better image quality and less computation time.

The proposed generator architecture we refer to as Trans-Net is shown in Fig. 2(a). It consists of a coding operation (left side) and a decoding operation (right side). The sizes and numbers of the feature channels in every layer are written in the side of both paths. Shortcuts are used to concatenate the features from the coding phase to the decoding phase in all corresponding downsampling and upsampling blocks as shown in Fig. 2(a). This can avoid the gradient vanishing or exploding problem during backpropagation, making it easy to train deep networks.

Discriminator. Since the L_1 or L_2 term can successfully capture the low-frequency information but fail to restore high-frequency information, producing blurred details on image generation tasks [13]. In order to generate both the low-frequency and the high-frequency details, we added a patch-level classifier as discriminator as proposed in [10]. This discriminator we refer to as PatchGAN can learn high-frequency features while the L_1 loss can learn low-frequency features. By fusing the two types of losses, both the high-frequency and the low-frequency details can be learnt and generated. PatchGAN restricts the attention to the structure in local image patches, which penalizes the structure at the scale of 70×70, aiming to classify whether 70×70 overlapping image patches are real or fake. The output of PatchGAN is a 30×30 matrix where every element have a receptive filed of 70×70, averaging the matrix to provide the final output.

Although patch is much smaller than the image, they can still generate high quality results with fewer parameters, and faster inference speed than that at image level. In addition, it can work on arbitrarily-sized images in a fully convolutional fashion [11]. Such a discriminator effectively models the image as a Markov random field, assuming independence between pixels separated by more than a patch diameter. The Markov random field characterize the image by local fragmentation region of pixel values. Therefore, our PatchGAN can be treated as a form of texture/style loss.

3   Experiments and Results

To have fair and comprehensive comparisons with other methods, we evaluated our model as follows: (i) Analysis of the image quality at different levels of downsampling; (ii) Analysis of the effect of the proposed generator on the results and comparisons of results using different generators; (iii) Quantitative and qualitative comparisons between our method and state-of-the-art approaches [20]. We will introduce the experimental dataset, the training details, the evaluational metrics and experimental results in the following sections.

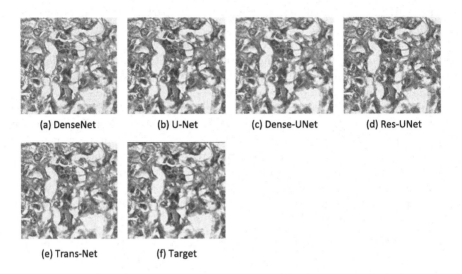

(a) DenseNet   (b) U-Net   (c) Dense-UNet   (d) Res-UNet

(e) Trans-Net   (f) Target

Fig. 3. Visual comparisons of results using different generators.

3.1   Dateset and Details

Dataset. The dataset is publicly available as part of the MITOS-ATYPIA14 challenge[1]. The dataset consists of 424 frames at $X20$ magnification which were stained with standard Hematoxylin and Eosin (H&E). The training dataset consists of 300 frames and the test dataset consists of 124 frames. All frames were scanned by two scanners: Aperio Scanscope XT and Hamamatsu Nanozoomer 2.0-HT. Slides from both scanners were resized to identical dimensions (1539 × 1376). For training, we extracted 9000 unpaired patches from the training dataset of both scanners. During evaluation, we randomly extracted 620 paired patches from the testing dataset of both scanners. All patches have the same size of 256 × 256. Non-rigid registration was employed to eliminate the mismatch. Patches from Scanner H were regarded as the ground truth.

Training Details. For all experiments, we trained 9000 unpaired patches from both scanners for 6 epochs with a batch size of 1. We used the Adam solver and trained all networks with a learning rate of 0.0002. We set $\lambda = 10$ in Eq. 4.

---

[1] https://mitos-atypia-14.grand-challenge.org.

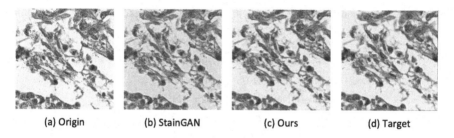

(a) Origin   (b) StainGAN   (c) Ours   (d) Target

Fig. 4. Visual comparison between the result of our proposed method and that of StainGAN.

Table 1. Results of Trans-Net at different levels: Mean indicators and total processing time

Methods	PSNR	SSIM	Time (sec)
Level8	22.22	0.812	31
Level7	21.67	0.793	29
Level6	21.99	0.807	28
Level5	21.97	0.806	24
Level4	21.99	0.802	22

We replaced the negative log likelihood objective for L_{GAN} by a least-squares loss [16] and updated the discriminators using a history of generated images rather than the ones produced by the latest generators. We kept an image buffer that stores the 50 previously created images. We used this image buffer which has information of previous 50 images rather than the latest image to update the discriminators. The hardware of GeForce GTX 1080 and the PyTorch framework were used.

Evaluation Metrics. Results were compared to the ground truth with two similarity metrics: Peak Signal-to-Noise Ration (PSNR) and Structural Similarity index (SSIM). In addition, the processing speed which is an important factor in clinical has been reported in the results. We used the total time over processing the 620 images from testing dataset to calculate the computational time.

3.2   Results with Different Levels of Downsampling

It is interesting to see some results demonstrate the image quality with different levels of downsampling. We increase the numbers of downsampling layer and upsampling layer from 4 to 8 symmetrically. Results is shown in Table 1. It is obviously that the result at sampling level of 8 is the best. And the 8 level is the highest level that we can sample, because the size of our input is $256 * 256$ and the size of the feature maps at this moment is $1 * 1$. Then, we adopt 8 downsampling layers and 8 upsampling layers for our generative networks.

Table 2. Comparison to other generators: Mean indicators and total processing time

Methods	PSNR	SSIM	Time (sec)
DenseNet	21.40	0.792	35
U-Net	21.95	0.802	45
Dense-UNet	20.10	0.795	119
Res-UNet	22.02	0.802	51
Trans-Net	**22.22**	**0.812**	**31**

Table 3. Stain Transfer Comparison: Mean indicators and total processing time

Methods	PSNR	SSIM	Time (sec)
StainGAN	21.35	0.785	60
Ours	**22.22**	**0.812**	31

3.3   Comparisons of Results Using Different Generators

The generator of our generative adversarial networks we refer to as Trans-Net is based on the traditional U-Net structure [19]. In order to make a extensive comparisons, we replaced the generator with other network structures and compared the results using our proposed Trans-Net and other generators.

Four other network structures were used as generators for comparison, including the traditional U-Net [19] which was used in biomedical image segmentation, Res-UNet [21] which was first proposed for segmentation of retinal images, the DenseNet structure which was used in classification tasks [8], and the Dense-UNet [7] which was used to remove artifacts from the image respectively. Results are shown in Table 2. The proposed Trans-Net achieved higher values than other generators in terms of PSNR and SSIM with less computational times. The visual comparisons as shown in Fig. 3, demonstrates that the results generated by the proposed Trans-Net are closer to the ground truth than the results using other generators in terms of colors, contrast and texture details.

3.4   Comparison with State-of-the-Art Method

The goal is to transfer the style of pathes from scanner A (Aperio) to the patches from scanner H (Hamamatsu) while keeping the context and texture of A. Since the medical images are unpaired in different centers, we used the cycle-consistent loss [22] to map the patches from domain A to domain H, and compared the generated patches with the real patches of scanner H (ground truth). The state-of-the-art method is StainGAN [20], the difference between TAN and StainGAN is that we have designed a novel generator we refer to as Trans-Net. The author of StainGAN adopted the architecture for their generative networks from Johnson et al. [11] who have shown impressive results for neural style transfer and super-resolution. This network contains two stride-2 convolutions, 6 residual blocks, and two stride-$\frac{1}{2}$ convolutions. Compared to the network, Trans-Net has much more stride-2 convolutions and stride-$\frac{1}{2}$ convolutions which results in more detailed semantic context information. And Trans-Net can produce much more detailed texture and color information owning to the operation that add the information before sampling to sampled information novelty. Finally, Trans-Net has only 16 convolutions, the reduction of convolutions accelerates the training speed and avoids the over-fitting problem. The results of the proposed method and StainGAN are shown in Table 3, and Fig. 4 shows their visual comparison. The PSNR value improves from 21.35 to 22.22 while SSIM improves from 0.785 to 0.812. In addition, the proposed method only requires half of the computational time of StainGAN. The visual comparison also show that the results generated by the proposed method are closer to the ground truth than the results using StainGAN.

4   Discussion and Conclusion

In this work, we proposed a novel method called TAN for stain style transfer. The experimental results show that the proposed method outperforms the state-of-the-art method in terms of objective metrics and visual comparisons. A new network structure called Trans-Net was proposed as generator, which contributes to better results than state-of-the-art results. There are three factors that contribute to the advantage of the Trans-Net structure: (1) It has many downsampling layers and upsampling layers to ensure the high-level semantic information can be learnt which can result in detailed texture and context. (2) It directly adds the information before and after sampling to reduce the loss of information which contributes to the much closer color and texture to the ground truth. (3) It only has 16 convolutional layers which accelerates the networks. It would be interesting to investigate how the stain style transfer affects the segmentation task and the analysis of pathological slides in our future work.

Acknowledgement. This study was supported by the National Natural Science Foundation of China (Grant No. 6190010435) and the Science and Technology Program of Fujian Province, China (Grant No. 2019YZ016006).

References

1. Basavanhally, A., Madabhushi, A.: EM-based segmentation-driven color standardization of digitized histopathology. In: Medical Imaging 2013: Digital Pathology, vol. 8676, p. 86760G. International Society for Optics and Photonics (2013)
2. Bautista, P.A., Hashimoto, N., Yagi, Y.: Color standardization in whole slide imaging using a color calibration slide. J. Pathol. Inform. **5**, 4 (2014)
3. Bejnordi, B.E., et al.: Stain specific standardization of whole-slide histopathological images. IEEE Trans. Med. Imaging **35**(2), 404–415 (2015)
4. Bejnordi, B.E., Timofeeva, N., Otte-Höller, I., Karssemeijer, N., van der Laak, J.A.: Quantitative analysis of stain variability in histology slides and an algorithm for standardization. In: Medical Imaging 2014: Digital Pathology, vol. 9041, p. 904108. International Society for Optics and Photonics (2014)

5. BenTaieb, A., Hamarneh, G.: Adversarial stain transfer for histopathology image analysis. IEEE Trans. Med. Imaging **37**(3), 792–802 (2017)
6. Ciompi, F., et al.: The importance of stain normalization in colorectal tissue classification with convolutional networks. In: 2017 IEEE 14th International Symposium on Biomedical Imaging (ISBI 2017), pp. 160–163. IEEE (2017)
7. Guan, S., Khan, A., Sikdar, S., Chitnis, P.: Fully dense UNet for 2D sparse photoacoustic tomography artifact removal. IEEE J. Biomed. Health Inform. (2019)
8. Huang, G., Liu, Z., Van Der Maaten, L., Weinberger, K.Q.: Densely connected convolutional networks. In: Proceedings of the IEEE Conference on Computer Vision and Pattern Recognition, pp. 4700–4708 (2017)
9. Ismail, S.M., et al.: Observer variation in histopathological diagnosis and grading of cervical intraepithelial neoplasia. BMJ **298**(6675), 707–710 (1989)
10. Isola, P., Zhu, J.Y., Zhou, T., Efros, A.A.: Image-to-image translation with conditional adversarial networks. In: Proceedings of the IEEE Conference on Computer Vision and Pattern Recognition, pp. 1125–1134 (2017)
11. Johnson, J., Alahi, A., Fei-Fei, L.: Perceptual losses for real-time style transfer and super-resolution. In: Leibe, B., Matas, J., Sebe, N., Welling, M. (eds.) ECCV 2016, Part II. LNCS, vol. 9906, pp. 694–711. Springer, Cham (2016). https://doi.org/10.1007/978-3-319-46475-6_43
12. Khan, A.M., Rajpoot, N., Treanor, D., Magee, D.: A nonlinear mapping approach to stain normalization in digital histopathology images using image-specific color deconvolution. IEEE Trans. Biomed. Eng. **61**(6), 1729–1738 (2014)
13. Larsen, A.B.L., Sønderby, S.K., Larochelle, H., Winther, O.: Autoencoding beyond pixels using a learned similarity metric. arXiv preprint: arXiv:1512.09300 (2015)
14. Macenko, M., et al.: A method for normalizing histology slides for quantitative analysis. In: 2009 IEEE International Symposium on Biomedical Imaging: From Nano to Macro, pp. 1107–1110. IEEE (2009)
15. Magee, D., et al.: Colour normalisation in digital histopathology images. In: Proceedings of Optical Tissue Image analysis in Microscopy, Histopathology and Endoscopy (MICCAI Workshop), vol. 100, pp. 100–111. Citeseer (2009)
16. Mao, X., Li, Q., Xie, H., Lau, R.Y., Wang, Z., Paul Smolley, S.: Least squares generative adversarial networks. In: Proceedings of the IEEE International Conference on Computer Vision, pp. 2794–2802 (2017)
17. Niethammer, M., Borland, D., Marron, J.S., Woosley, J., Thomas, N.E.: Appearance normalization of histology slides. In: Wang, F., Yan, P., Suzuki, K., Shen, D. (eds.) MLMI 2010. LNCS, vol. 6357, pp. 58–66. Springer, Heidelberg (2010). https://doi.org/10.1007/978-3-642-15948-0_8
18. Reinhard, E., Adhikhmin, M., Gooch, B., Shirley, P.: Color transfer between images. IEEE Comput. Graph. Appl. **21**(5), 34–41 (2001)
19. Ronneberger, O., Fischer, P., Brox, T.: U-Net: convolutional networks for biomedical image segmentation. In: Navab, N., Hornegger, J., Wells, W.M., Frangi, A.F. (eds.) MICCAI 2015, Part III. LNCS, vol. 9351, pp. 234–241. Springer, Cham (2015). https://doi.org/10.1007/978-3-319-24574-4_28
20. Shaban, M.T., Baur, C., Navab, N., Albarqouni, S.: StainGAN: stain style transfer for digital histological images. arXiv preprint: arXiv:1804.01601 (2018)
21. Xiao, X., Lian, S., Luo, Z., Li, S.: Weighted Res-UNet for high-quality retina vessel segmentation. In: 2018 9th International Conference on Information Technology in Medicine and Education (ITME), pp. 327–331. IEEE (2018)
22. Zhu, J.Y., Park, T., Isola, P., Efros, A.A.: Unpaired image-to-image translation using cycle-consistent adversarial networks. In: Proceedings of the IEEE International Conference on Computer Vision, pp. 2223–2232 (2017)

Blind Deconvolution Microscopy
Using Cycle Consistent CNN
with Explicit PSF Layer

Sungjun Lim[1] and Jong Chul Ye[2(✉)]

[1] KAIST Institute for Artificial Intelligence, Daejeon, Republic of Korea
[2] Department of Bio/Brain Engineering, KAIST, Daejeon, Republic of Korea
jong.ye@kaist.ac.kr

Abstract. Deconvolution microscopy has been extensively used to improve the resolution of the widefield fluorescent microscopy. Conventional approaches, which usually require the point spread function (PSF) measurement or blind estimation, are however computationally expensive. Recently, CNN based approaches have been explored as a fast and high performance alternative. In this paper, we present a novel unsupervised deep neural network for blind deconvolution based on cycle consistency and PSF modeling layers. In contrast to the recent CNN approaches for similar problem, the explicit PSF modeling layers improve the robustness of the algorithm. Experimental results confirm the efficacy of the algorithm.

Keywords: Microscopy · Image reconstruction · Machine learning

1 Introduction

In fluorescent microscopy, light diffraction from a given optics degrades the resolution of images. To improve resolution, many optimization-based deconvolution algorithms have been developed [2,12,16]. When the PSF measurements are not available, You et al. [19] proposed a blind deconvolution method by solving joint minimization problem to estimate the unknown blur kernel and the image. Chan et al. [1] proposed an improved version of blind deconvolution using TV regularization.

Recently, convolutional neural networks (CNN) have been extensively used to enhance performance of an optical microscope without hardware changes. Rivenson et al. [15] used deep neural networks to improve optical microscopy, enhancing its spatial resolution over a large field of view and depth of field. Nehme et al. [13] used deep convolutional neural network that can be trained on simulated data or experimental measurement to obtain super resolution images from localization microscopy. Weigert et al. [18] proposed CNN method which can recover isotropic resolution from anisotropic data. In addition, generative adversarial network (GAN) has attracted much attention in inverse problem by providing a way to use unlabeled data to train a deep neural network [11]. Kupyn et al. [7] presented DeblurGAN for motion deblurring using a conditional GAN and content loss. However, this GAN approaches often generates the artificial features due to the mode collapsing, so a cycle-consistent adversarial network (Cycle-GAN) [20] that imposes the one-to-one correspondency has also made impact on image reconstruction [6,8].

© Springer Nature Switzerland AG 2019
F. Knoll et al. (Eds.): MLMIR 2019, LNCS 11905, pp. 173–180, 2019.
https://doi.org/10.1007/978-3-030-33843-5_16

However, these CycleGAN approaches usually require two generators with high capacity, which are often difficult to train with small number of training data. To address this problem, this paper proposes a novel CycleGAN architecture with an explicit PSF layer for blind deconvolution problems. Thanks to the simple PSF layer that generates blur images, we show that our proposed method is robust and efficient for the deconvolution task in spite of fully exploiting the cyclic consistency for blind deconvolution.

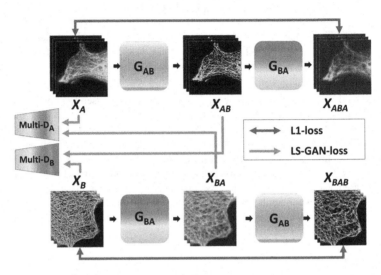

Fig. 1. Overall architecture of our proposed method. G_{AB} are generators that map the blur domain to the sharp image domain, and G_{BA} is an explicit PSF layer that needs to be estimated. Multi-D_A,D_B are modules that contain independent discriminators which take cropped patches on different scale.

2   Theory

Figure 1 illustrates overall framework of the proposed method. We refer to A as the blurred image domain and B as the blur-removed sharp image domain. The generator G_{AB} then maps a blurred image in A to a sharp image in B, and the generator G_{BA} corresponds to blur operation from sharp image domain B to a blurred measurement domain A. In contrast to the existing cycle-GAN architecture for blind deconvolution [8], we use an explicit PSF layer for the map G_{BA}, in which the actual PSF values are estimated from the training data.

While the use of an explicit PSF layer can have a risk to reduce the generalizability of the PSF, we found that in typical microscopic setups with predetermined optics, the PSF is generally fixed so that sample-dependent PSF adaptation is not much necessary. Instead, the use of an explicit PSF layer significantly improves the stability of the algorithm.

In addition, the discriminator network D_A is designed to distinguish the synthetically generated blurred image from real ones. Similarly, D_B is to discriminate generated deblurred images from sharp image distribution. For the sharp image distribution, we use *un-matched* high resolution images. These could come from super-resolution microscopy or from commercially available deconvolution software. Finally, we train both the generators and the discriminators in an alternating manner by solving the following optimization problem:

$$\min_{G_{AB}, G_{BA}} \max_{D_A, D_B} L(G_{AB}, G_{BA}, D_A, D_B) \tag{1}$$

in which the loss function is defined as follows:

$$L(G_{AB}, G_{BA}, D_A, D_B) = L_{GAN}(G_{AB}, D_B, A, B) + L_{GAN}(G_{BA}, D_A, B, A)$$
$$+ \lambda_1 L_{cyclic}(G_{AB}, G_{BA}) + \lambda_2 \|G_{BA}\|_1$$

where λ_1, λ_2 are hyperparameters, and L_{GAN}, L_{Cyclic} are an adversarial loss, cyclic loss respectively. $\|G_{BA}\|_1$ is the L1-norm for the regularization of blur kernel. In following sections, we will give further explanation regarding each component of the loss function.

2.1 Loss Function

Adversarial Loss. We employed the modified GAN loss using a Least Square Loss [10]. Specifically, the min-max optimization problem for GAN training is composed of two separate minimization problems as follows:

$$\min_{G_{AB}} \mathbb{E}_{x_A \sim P_A} \left[(D_B(G_{AB}(x_A)) - 1)^2 \right] \tag{2}$$

$$\min_{D_B} \frac{1}{2} \mathbb{E}_{x_B \sim P_B} \left[(D_B(x_B) - 1)^2 \right] + \frac{1}{2} \mathbb{E}_{x_A \sim P_A} \left[D_B(G_{AB}(x_A))^2 \right] \tag{3}$$

where P_A and P_B denote the distribution for the domain A and B. By optimizing the adversarial loss, we can regulate the generators so that the generated sharp image volume is as realistic as possible; at the same time, the discriminators are optimized to distinguish the generated deconvoluted image volume from the real high resolution image. The equivalent adversarial loss was also imposed on G_{BA} for deceiving generation of synthetic blurred data.

Cyclic Loss. Although mapping between (A) and (B) can be estimated by a well trained adversarial network, it is still vulnerable to the mode failure problem in which many input images are taken into a fixed output image. Also, because of the large capacity of a deep neural network, the network can map (A) to any random permutation of the output in the domain (B) that the target distribution is likely to match. In other words, the adversarial loss alone cannot guarantee a reversal of both domains. In order to resolve such issues, Zhu et al. [20] proposed cycle consistency loss. In our case, the loss of cycle consistency supports a one-to-one correspondence between the blurred image volume and the deconvoluted volume. The specific cycle consistency loss is defined as follows:

$$L_{cyclic}(G_{AB}, G_{BA}) = \mathbb{E}_{x_A \sim P_A} \left[\|G_{BA}(G_{AB}(x_A)) - x_A\|_1 \right] + \mathbb{E}_{x_B \sim P_B} \left[\|G_{AB}(G_{BA}(x_B)) - x_B\|_1 \right] \tag{4}$$

2.2   Multi patchGANs in CycleGAN

As for the discriminators, we propose an improved model from the original CycleGAN using multi-PatchGANs (mPGANs) [5], where each discriminator has input patches with different sizes used. PatchGAN typically focuses on high-frequency structures by including local patches for the entire image. Because patches with different scales can contain different high-frequency structures, we use multiple discriminators that take the patches at different scales. Specifically, we define multi-discriminator as $\{D_A^{f_i}, D_B^{f_i}\}$ where f_i denotes the ith scale patch. The adversarial loss with the multiscale patches is then formulated as follows:

$$L_{GAN}(G_{AB}, D_B, A, B) = \mathbb{E}_{X_B \sim P_B}\left[\sum_{i=1}^{N}\left(1 - D_B^{f_i}(X_B)\right)^2\right] + \mathbb{E}_{X_A \sim P_A}\left[\sum_{i=1}^{N}\left(D_B^{f_i}(G_{AB}(X_A))\right)^2\right]$$

(5)

where N-denotes the number of total scales. $L_{GAN}(G_{BA}, D_A, B, A)$ is similarly defined.

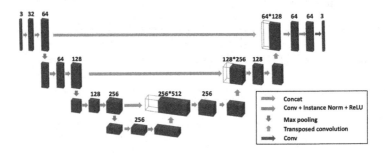

Fig. 2. 3D U-net architecture for our generator.

Fig. 3. 3D discriminator architecture. The discriminator consists of 4 modules which consist of Conv + Instance Norm + ReLU. Every Conv layer has stride 2, and downsamples the input volume. At last layer, the number of output channel is 1.

3   Network Architecture

The network architecture of the generator G_{AB} is 3D-Unet [3] as illustrated in Fig. 2. For the architecture of G_{AB}, our U-net structure consists of contracting, expanding paths. The contracting path consists of the repetition of the following blocks: 3D conv-Instance Normalization [17]- ReLU. Through the network, the convolutional kernel

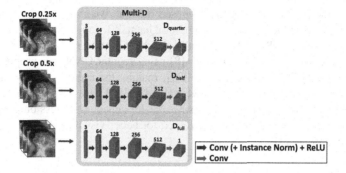

Fig. 4. Our multiple discriminators consist of three independent discriminators. Each discriminator takes patches at different scales. Specifically, D_{full} takes the patch in its original size; D_{half} takes the randomly cropped patch half size of the original patch size; and $D_{quarter}$ takes the randomly cropped patch a quarter of the original patch size.

dimension is $3 \times 3 \times 3$. At the first layer, a channel of the feature map is 64. The network architecture of discriminators $D_A^{f_i}, D_B^{f_i}$ is illustrated in Fig. 3. The discriminators are PatchGANs [5], and we use 3 discriminators that process patches with 3 different scales as shown in Fig. 4. The network architecture of the discriminators consist of modules, which consist of 3D conv- Instance Normalization- ReLU. Through the network, the convolution kernel dimension is $3 \times 3 \times 3$.

On the other hand, the generator G_{BA} uses only a single 3D convolution layer to model a 3D blurring kernel. The size of the 3D PSF modeling layer is chosen depending on training set.

4   Method

For training, we used 19 epifluorescence (EPF) samples of tubulin with a size of $512 \times 512 \times 30$. As for unmatched sharp image volume, we use deblurred image generated by utilizing a commercial software AutoQuant X3 (Media Cybernetics, Rockville). The volume depth is increased to 64 by padding with reflect. Due to memory limitations, the volume is split into $64 \times 64 \times 64$ patches. For data augmentation, rotation, flip, translation, and scale are imposed on the input patches. Adam optimizer with $\beta_1 = 0.9$ and $\beta_2 = 0.999$ is used to optimize the Eq. (1), and the learning rate is 0.0001. The learning rate decreases linearly after epoch 40; and the total number of epoch is 200. To reduce model oscillation [4], the discriminators used a history of generated volumes from a frame buffer containing 50 previously generated volumes. For all experiments, we set λ_1 of (1) as 3 and λ_2 as 0.01. For the optimizer, we used only a single batch. We normalized the patches and set them to [0,1]. The PSF size is set to 20. The proposed method was implemented in Python with Tensorflow, and GeForce GTX 1080 Ti GPU was used for both training and testing the network.

To verify the performance of the proposed method, we compare our method with commercial deconvolution method using AutoQuant X3, supervised learning [9], and the original cycleGAN [8] with both multi-PatchGANs and G_{BA} from another CNN (Non-PSF layer). In contrast to Lu et al. [8] using regular CNN, our proposed model only used single PSF modeling layer in G_{BA}, making the training process much easier. For supervised learning network, we trained a 3D U-net with the matched label data from AutoQuant X3 using L1-loss since L1-loss encourages less blurring [5]. All the reconstruction results were post-processed for better visualization by adaptive histogram equalization [14].

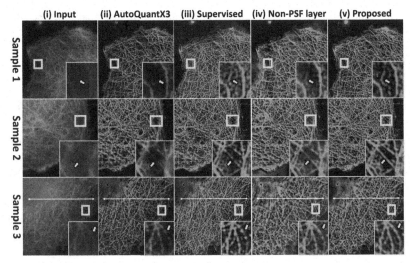

(a) Comparison of 3 test samples of transverse view over the follwing methods: AutoQuantX3, supervised learning, CycleGAN with both multi-PatchGANs and non-PSF layer (Non-PSF layer), and the proposed method. ROI (marked yellow) in lower right corner shows enlarged result.

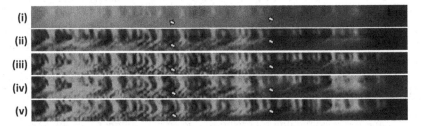

(b) Sagittal views of Sample 3 from Fig. 5a. The marked white line on sample 3 from Fig. 5a shows the scan line of the presented sagittal views.

Fig. 5. Result of transverse view and sagittal view.

5   Experimental Results

Figure 5a and b show cross-views and sagittal views of various reconstruction method. In Fig. 5a, input images are degraded by blur and noise. Besides, as shown in Fig. 5b, the degradation at deeper depth gets worse. In Fig. 5a, AutoQuant X3 removed blur and noise; however, it did not improve contrast sufficiently. Both the supervised learning and the non-PSF layer showed better contrast and removed blur; however, the structural continuity was not preserved. In Fig. 5b, the AutoQuant X3, the supervised learning, and the non-PSF layer somehow removed blur and noise, but did not maintain structure continuity at deeper depth. Finally, in the proposed method blurs and noise were successfully removed in both Fig. 5a and b, thereby preserving the continuity of the structure. We therefore confirm that PSF modeling layer improves the robustness of the proposed method.

6   Discussion and Conclusion

In this paper, we presented a novel blind deconvolution using an unsupervised deep neural network using CycleGAN architecture. Experimental results showed that our proposed method restores the good quality reconstruction in both transverse and sagittal view. In particular, we observed that the use of PSF modeling layer improved the effectiveness of the proposed method. We have also proposed multiple patchGANs taking patches at different scales to discriminate real samples from generated results. The multiple patchGANs helped generators to produce coarsest and finest details.

References

1. Chan, T.F., Wong, C.K.: Total variation blind deconvolution. IEEE Trans. Image Process. **7**(3), 370–375 (1998)
2. Chaudhuri, S., Velmurugan, R., Rameshan, R.: Blind deconvolution methods: a review. In: Blind Image Deconvolution: Methods and Convergence, pp. 37–60. Springer, Cham (2014). https://doi.org/10.1007/978-3-319-10485-0_3
3. Çiçek, Ö., Abdulkadir, A., Lienkamp, S.S., Brox, T., Ronneberger, O.: 3D U-Net: learning dense volumetric segmentation from sparse annotation. In: Ourselin, S., Joskowicz, L., Sabuncu, M.R., Unal, G., Wells, W. (eds.) MICCAI 2016, Part II. LNCS, vol. 9901, pp. 424–432. Springer, Cham (2016). https://doi.org/10.1007/978-3-319-46723-8_49
4. Goodfellow, I., et al.: Generative adversarial nets. In: Advances in Neural Information Processing Systems, pp. 2672–2680 (2014)
5. Isola, P., Zhu, J.Y., Zhou, T., Efros, A.A.: Image-to-image translation with conditional adversarial networks. In: Proceedings of the IEEE Conference on Computer Vision and Pattern Recognition, pp. 1125–1134 (2017)
6. Kang, E., Koo, H.J., Yang, D.H., Seo, J.B., Ye, J.C.: Cycle-consistent adversarial denoising network for multiphase coronary CT angiography. Med. Phys. **46**(2), 550–562 (2019)
7. Kupyn, O., Budzan, V., Mykhailych, M., Mishkin, D., Matas, J.: DeblurGAN: blind motion deblurring using conditional adversarial networks. In: Proceedings of the IEEE Conference on Computer Vision and Pattern Recognition, pp. 8183–8192 (2018)
8. Lu, Y., Tai, Y.W., Tang, C.K.: Conditional CycleGAN for attribute guided face image generation. arXiv preprint: arXiv:1705.09966 (2017)

9. Mao, X., Shen, C., Yang, Y.B.: Image restoration using very deep convolutional encoder-decoder networks with symmetric skip connections. In: Advances in Neural Information Processing Systems, pp. 2802–2810 (2016)
10. Mao, X., Li, Q., Xie, H., Lau, R.Y., Wang, Z., Paul Smolley, S.: Least squares generative adversarial networks. In: Proceedings of the IEEE International Conference on Computer Vision, pp. 2794–2802 (2017)
11. McCann, M.T., Jin, K.H., Unser, M.: Convolutional neural networks for inverse problems in imaging: a review. IEEE Signal Process. Mag. **34**(6), 85–95 (2017)
12. McNally, J.G., Karpova, T., Cooper, J., Conchello, J.A.: Three-dimensional imaging by deconvolution microscopy. Methods **19**(3), 373–385 (1999)
13. Nehme, E., Weiss, L.E., Michaeli, T., Shechtman, Y.: Deep-storm: super-resolution single-molecule microscopy by deep learning. Optica **5**(4), 458–464 (2018)
14. Pizer, S.M., et al.: Adaptive histogram equalization and its variations. Comput. Vis. Graph. Image Process. **39**(3), 355–368 (1987)
15. Rivenson, Y., Göröcs, Z., Günaydin, H., Zhang, Y., Wang, H., Ozcan, A.: Deep learning microscopy. Optica **4**(11), 1437–1443 (2017)
16. Sarder, P., Nehorai, A.: Deconvolution methods for 3-D fluorescence microscopy images. IEEE Signal Process. Mag. **23**(3), 32–45 (2006)
17. Ulyanov, D., Vedaldi, A., Lempitsky, V.: Instance normalization: the missing ingredient for fast stylization. arXiv preprint: arXiv:1607.08022 (2016)
18. Weigert, M., Royer, L., Jug, F., Myers, G.: Isotropic reconstruction of 3D fluorescence microscopy images using convolutional neural networks. In: Descoteaux, M., Maier-Hein, L., Franz, A., Jannin, P., Collins, D.L., Duchesne, S. (eds.) MICCAI 2017, Part II. LNCS, vol. 10434, pp. 126–134. Springer, Cham (2017). https://doi.org/10.1007/978-3-319-66185-8_15
19. You, Y.L., Kaveh, M.: A regularization approach to joint blur identification and image restoration. IEEE Trans. Image Process. **5**(3), 416–428 (1996)
20. Zhu, J.Y., Park, T., Isola, P., Efros, A.A.: Unpaired image-to-image translation using cycle-consistent adversarial networks. In: Proceedings of the IEEE International Conference on Computer Vision, pp. 2223–2232 (2017)

Deep Learning Based Approach to Quantification of PET Tracer Uptake in Small Tumors

Laura Dal Toso[1(✉)], Elisabeth Pfaehler[2], Ronald Boellaard[2,3],
Julia A. Schnabel[1], and Paul K. Marsden[1]

[1] School of Biomedical Engineering and Imaging Sciences,
King's College London, London, UK
laura.dal_toso@kcl.ac.uk
[2] Department of Nuclear Medicine and Molecular Imaging,
University of Groningen, University Medical Center Groningen,
Groningen, The Netherlands
[3] Department of Radiology and Nuclear Medicine,
Amsterdam University Medical Centers, Amsterdam, The Netherlands

Abstract. In Positron Emission Tomography (PET), quantification of tumor radiotracer uptake is mainly performed using standardised uptake value and related methods. However, the accuracy of these metrics is limited by the poor spatial resolution and noise properties of PET images. Therefore, there is a great need for new methods that allow for accurate and reproducible quantification of tumor radiotracer uptake, particularly for small regions. In this work, we propose a deep learning approach to improve quantification of PET tracer uptake in small tumors using a 3D convolutional neural network. The network was trained on simulated images that present 3D shapes with typical tumor tracer uptake distributions ('ground truth distributions'), and the corresponding set of simulated PET images. The network was tested on unseen simulated PET images and was shown to robustly estimate the original radiotracer uptake, yielding improved images both in terms of shape and activity distribution. The same network was successful when applied to 3D tumors acquired from physical phantom PET scans.

Keywords: Convolutional neural network · PET · Quantification · Reconstruction

1 Introduction

1.1 Positron Emission Tomography

Positron Emission Tomography (PET) is widely used in clinical oncology for the evaluation of lesion malignancy, staging and for monitoring the tumor response

This project has received funding from the European Union's Horizon 2020 research and innovation programme under the Marie Skłodowska-Curie grant agreement No. 764458.

© Springer Nature Switzerland AG 2019
F. Knoll et al. (Eds.): MLMIR 2019, LNCS 11905, pp. 181–192, 2019.
https://doi.org/10.1007/978-3-030-33843-5_17

to treatment [11]. In the clinical routine, images are often interpreted by visual inspection, together with semi-quantitative measurements of tumor radiotracer uptake such as standardized uptake value (SUV) and related metrics. The SUV is defined as follows:

$$SUV = \frac{activity\ concentration\ in\ ROI}{average\ activity\ concentration\ in\ whole\ body} \tag{1}$$

Two common ways of reporting SUV are SUV_{max} and SUV_{mean}. SUV_{max} represents the highest voxel value in a region of interest (ROI). This measurement is insensitive to the tumor boundary definition but it is very susceptible to noise. SUV_{mean} is an average SUV calculated over voxels in a boundary ROI. As a result, it is less sensitive to noise but it is dependent on the ROI definition and it typically has a lower value than SUV_{max}. An alternative metric is SUV_{peak} which is an average SUV calculated inside a small ROI, usually a 1 mL spherical volume, containing the pixel with maximum intensity [13]. SUV_{peak} is less affected by noise but depends on the ROI's shape, size and location. It is complicated to accurately quantify tumor uptake in PET images due mainly to poor spatial resolution, typically 5 mm FWHM, and noise [1,12]. Important advances have recently been made in the development of techniques such as tumor segmentation and image reconstruction, but there is a great need for new accurate and reproducible quantification methods and that can be easily integrated in clinical and clinical research settings [2].

1.2   Deep Learning in PET Imaging

In recent years deep learning techniques have been massively applied to medical imaging. These approaches have been extremely successful in performing tasks such as segmentation, classification, automatic detection and, to a lesser extent so far, image reconstruction [7,8]. Convolutional neural networks (CNNs) have been successfully applied to PET images to perform denoising [5], lesion detection and lesion segmentation [3], as well as image reconstruction [4]. Even though a lot of progress has been made in these areas, there are only few deep learning applications explicitly aimed at improving quantification in PET imaging. In this work we present a deep learning approach with the aim of more accurately quantifying tumor radiotracer uptake in PET studies.

2   Materials and Method

One of the main obstacles that hamper the application of deep learning to PET imaging is the lack of large labelled image datasets, that are needed to train the networks. Because of the difficulty in obtaining ground truth radiotracer distribution data, in this work a simulation algorithm was developed to generate synthetic datasets that were used to train and test the network. A 3D CNN was trained on simulated PET images and on the corresponding ground truth radiotracer distributions. The 3D CNN learned the relationship between the two sets of images and, when presented with an unseen set of simulated PET images, it restored an estimate of the true radioactivity distribution. The proposed method, illustrated in Fig. 1, will be described in the following paragraphs.

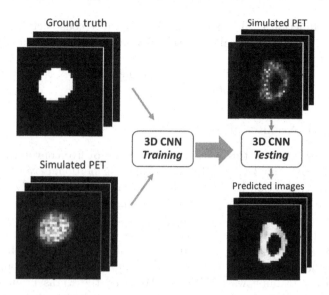

Fig. 1. A 3D CNN is trained on ground truth images and simulated PET images. The network is tested on an unseen set of simulated PET and it yields a prediction of the corresponding ground truth.

2.1 Generation of 3D Shapes and Radionuclide Distribution

Two sets of data, henceforth called ground truth images and simulated PET images, were generated using a simulation algorithm. The set of ground truth images was composed of:

- subset 1: 1400 warped spheres filled with uniform activity
- subset 2: 1400 warped spheres divided in two halves. Different values of uniform activity were assigned to each volume, with a ratio of 3:1
- subset 3: 1400 hollow warped spheres. Different values of uniform activity were assigned to each volume, the activity in the inner part being one third of the activity in the outer part

The background of each image was set to $1/10^{th}$ of the maximum activity. These specific patterns of radiotracer distribution were chosen to simulate realistic heterogeneous tumor uptake distributions [10]. The simulated images were made of $35 \times 35 \times 40$ voxels and the voxel size was set to $3.18 \times 3.18 \times 2.00$ mm^3. In each subset the radius of the 3D shapes before the warping process spanned 2 to 12 voxels (6 to 36 mm) and the activity concentrations spanned 2 kBq/mL to 50 kBq/mL. After generating the ground truth images, the corresponding simulated PET images were produced by applying Gaussian convolution (FWHM varying between 4 and 6 mm as described below), to simulate the effects of the point spread function (PSF) of the acquisition and reconstruction system and by adding noise. The noise values, assigned voxel by voxel, were drawn from a Gaussian distribution with $\sigma = k * \sqrt{N}$, where N is the number of counts in each voxel and k is a constant, set such that the noise level is equivalent to the noise observed in real PET data.

2.2 Network Architecture

The 3D network used in this work, was composed of five convolutional layers, each with 32 filters and $3 \times 3 \times 3$ filter size. Each convolutional layer was followed by a batch normalization layer to stabilize and accelerate the network training [6]. Finally, a fully connected layer with one hidden unit was used to obtain the output images. ReLu activation functions were assigned to the convolutional layers and a linear activation was used for the fully connected layer. The loss function was a mean squared error function, calculated between the predicted images and the ground truth images. The optimizer used to minimize the loss function during training was RMSprop. We used the Keras Framework with Tensorflow backend to implement the network, and training was performed on a NVIDIA Quadro M1200 GPU. Due to the limited available memory on the GPU, a batch size of 26 was used. The validation loss was monitored during the training process. The learning rate was set to the default value 0.001.

2.3 Testing the Procedure

Images were visually inspected, using a software tool for multimodality medical image analysis (AMIDE) [9], as well as quantitatively assessed. When the ground truth was available, the predicted images were assessed by calculating the mean recovery coefficient RC_{mean}, defined as a ratio of the mean intensity value $\bar{I}$ of each prediction and its corresponding ground truth:

$$RC_{mean} = \frac{\bar{I}_{prediction}}{\bar{I}_{ground\ truth}} \tag{2}$$

The mean value for a given image was calculated over the voxels that exceeded a threshold of 50% of the maximum intensity. The mean structural similarity (MSSIM) [14] was also calculated to evaluate the similarity between the predicted images and the ground truth images. Three sets of experiments were performed to assess the performance of the procedure and the impact of different parameters on the CNN's predictions.

Normalization
When deep networks are used to perform tasks like segmentation or classification, it is common practice to normalize the input data to make the training faster.

In our case the main goal is to improve quantification, so a study was performed to verify that absolute values are preserved during the normalization process. The dataset generated for this experiment, called original dataset, was made of 4200 images, divided into three subsets as described in Sect. 2.1. The simulated PET images corresponding to the first, second and third subset were produced using a Gaussian function with $FWHM = [4, 4, 4]$ mm, $FWHM = [5, 5, 5]$ mm and $FWHM = [6, 6, 6]$ mm respectively. To asses the impact of normalization, the original dataset was normalized by calculating the global maximum within the total set of simulated PET images and the total set of ground truth images, and by dividing each voxel in all images by the global maximum. By using this method, a single scaling factor was used for all images so it was possible to easily rescale the predicted images to the original units. The CNN was separately trained on 3460 images belonging to the non-normalized dataset and on 3460 images taken from normalized dataset and the results were compared.

Different Spatial Resolutions
One of the aims of this work is to generate a model that can be easily applied in the clinical routine, so ideally it should not depend on the properties (in particular the PSF) of any specific scanner. At first, we trained and tested the CNN on a dataset including simulated PET images generated using the same PSF. Then, the same CNN was tested on simulated PET images produced using different PSFs. The training dataset generated for this experiment was made of 1120 images, that consisted of warped spheres filled with uniform activity distributions. The simulated PET images in the training set were created using a Gaussian function with $FWHM = [4, 4, 4]$ mm. The first test set (TS1) was formed of 280 images generated in the same way. A second test set (TS2) was then created, using Gaussian functions with $FWHM = [6, 6, 6]$ mm to generate the simulated PET images. In this experiment the images used for training and testing were not normalized.

Physical Phantom PET Scans
After training and testing the CNN on simulated images, we have also tested the network on a small set of phantom data acquired on a PET scanner. In this case, the network was trained on the same training dataset used to test the effects of normalization, where the simulated PET images belonging to the first, second and third subset were produced using a Gaussian function with $FWHM = [4, 4, 4]$ mm, $FWHM = [5, 5, 5]$ mm and $FWHM = [6, 6, 6]$ mm respectively. The network was trained on 80% of the data belonging to the combined datasets made of three subsets, and tested on the remaining 20%. A fraction of 20% of the training set was used for validation. Then the network was tested on 3D patches extracted from real phantom PET scans. The physical phantom had the same size of a NEMA NU 2-2012 IQ phantom, that has a shape similar to a torso. Three 3D printed inserts simulating heterogeneous

uptake distributions and realistic tumor shapes [10] were placed in the phantom at equal distances. The first tumor insert (T1) had a volume of 46.00 mL and was filled with an activity solution of 19.49 kBq/mL. The second tumor (T2) was divided into two parts: the upper part (10.75 mL) filled with an activity solution of 10.94 kBq/mL and the lower part (13.12 mL) filled with an activity solution of 19.49 kBq/mL. The third tumor (T3) was hollow, the outer part (65.35 mL) filled with an activity solution of 19.49 kBq/mL and the inner core (7.80 mL) filled with non-radioactive water. The background compartment of the NEMA IQ phantom was filled with an activity solution of 1.94 kBq/mL. The phantom was scanned on a PET/CT system (Biograph mCT-40 PET/CT, Siemens, Knoxville, TN, USA) and the scans were acquired as list-mode data. The data were reconstructed to obtain a frame of 300 s, that is comparable to the scan time used in the clinic for patients, using an iterative ordered subset expectation maximization (OSEM) algorithm (3 iterations, 24 subsets), time-of-flight (TOF) iterative reconstruction (3 iterations, 21 subsets) and point spread function (PSF) modeling. The size of the 3D patches used to test the 3D CNN was $35 \times 35 \times 40$ voxels and each patch contained the image of one tumor insert.

3   Results

3.1   Normalization

The predicted images obtained for two representative volumes are shown in Fig. 2. The first column shows a coronal section of the ground truth for each volume. The second and third column show the corresponding predictions obtained training and testing the CNN on the normalized dataset and on the non-normalized dataset respectively. The images in Fig. 2(c) and (f) are more similar to the ground truth: the edges of the active volumes are more clearly defined and the predicted intensities are closer to the ground truth activity distributions. This visual assessment is supported by the RC_{mean} values and by the mean structural similarity shown in Fig. 3. These graphs confirm that the images predicted by the CNN tested on the non-normalized dataset are overall characterized by higher MSSIM and RC_{mean} values, meaning that they are overall more similar to the ground truth. For this reason, non-normalized data have been used in the subsequent experiments.

3.2   Different Spatial Resolutions

In Fig. 4 three transverse views of two representative volumes belonging respectively to TS1 (top row) and TS2 (bottom row) are shown. The CNN yields better predictions when tested on TS1, in which the simulated PET images were produced using the same Gaussian function as in the training set. By visual comparison we can notice that the prediction in Fig. 4(c) looks more similar to the ground truth in Fig. 4(a) both in terms of shape and activity distribution.

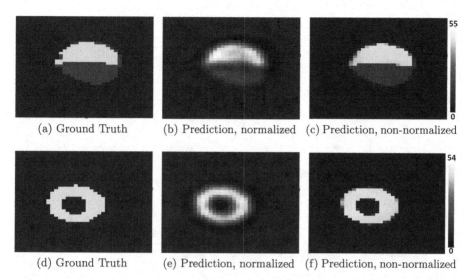

(a) Ground Truth   (b) Prediction, normalized   (c) Prediction, non-normalized

(d) Ground Truth   (e) Prediction, normalized   (f) Prediction, non-normalized

Fig. 2. Each row shows three coronal views, belonging to two representative volumes. The ground truth images are shown in the first column. The predicted images obtained testing the CNN on the normalized dataset, rescaled to the original units, are shown in the second column. The results obtained testing the CNN on the non-normalized dataset are shown in the third column. The network performs more effectively if the training data is not normalized. The intensities are expressed in kBq/mL.

The prediction in Fig. 4(f) has an overall lower activity and blurrier edges than the corresponding ground truth in Fig. 4(d). The results obtained for 3D shapes smaller than 40 mL, presented in Fig. 5, show that the CNN better recovers the mean intensity in the warped spheres belonging to the first test set, which has the same PSF as the training data. This indicates the importance of matching the PSFs in the training set and in the testing set.

3.3 Physical Phantom PET Scans

A fraction of 20% of the simulated PET images belonging to the combined dataset (made of three subsets), that had not been used for training, was used at first to test the CNN. The results obtained in this case showed that the CNN could recover well the ground truth activity distributions and shapes. Then the same network was tested on phantom data, the results obtained testing the network on 3D images of the phantom inserts T1 and T2 are shown in Fig. 6. Three images, reconstructed using OSEM, PSF and PSF+TOF are shown for each tumor. Directly under each phantom image, the corresponding CNN's prediction is presented. The images yielded by the CNN are less noisy and the edges of the tumors are better defined. Due to the lack of ground truth images, in this

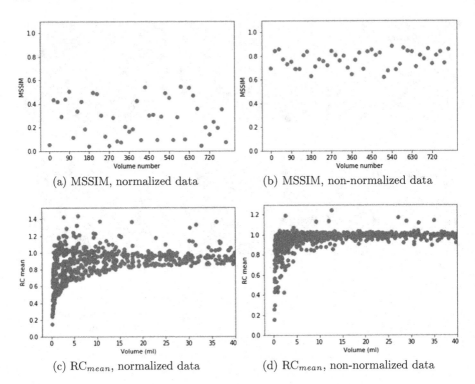

(a) MSSIM, normalized data

(b) MSSIM, non-normalized data

(c) RC_{mean}, normalized data

(d) RC_{mean}, non-normalized data

Fig. 3. Representation of MSSIM and RC_{mean} values, calculated for the predicted volumes belonging to the normalized dataset on the left and to the non-normalized dataset on the right. Only 45 representative MSSIM values are plotted to allow for a better visualization.

case it was not possible to estimate the MSSIM and RC_{mean}. The quantification was performed estimating the maximum intensity voxel in each volume, a measurement that can be related to SUV_{max}. The maximum values extracted from the real phantom images and from the predicted images are presented in Table 1. The ground truth maximum value is 19.49 kBq/mL for all tumor inserts. The predicted maximum values range from 18.18 kBq/mL to 23.98 kBq/mL, and are closer to the ground truth than the ones calculated for the real phantom scans. Although the CNN has a denoising effect, some noise is still present in the predicted images, which explains the variation observed in the predicted maximum values.

Training data 4mm PSF, test data 4mm PSF

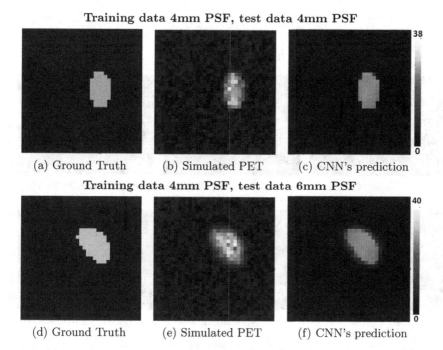

   (a) Ground Truth   (b) Simulated PET   (c) CNN's prediction

Training data 4mm PSF, test data 6mm PSF

   (d) Ground Truth   (e) Simulated PET   (f) CNN's prediction

Fig. 4. Illustration of three transverse views of two representative volumes, belonging to TS1 (top row) and to TS2 (bottom row) respectively. The ground truth images are presented in the first column. The corresponding simulated PET images, generated using a PSF with FWHM = [4, 4, 4] mm and FWHM = [6, 6, 6] mm are shown in the second column, in Fig. (b) and (e) respectively. The predicted images yielded by the CNN are shown in the third column. The intensities are expressed in kBq/mL.

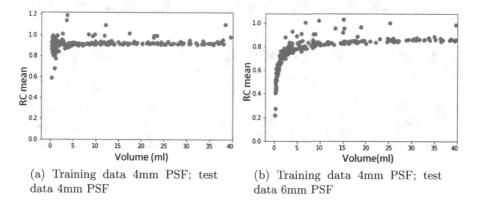

  (a) Training data 4mm PSF; test   (b) Training data 4mm PSF; test
  data 4mm PSF   data 6mm PSF

Fig. 5. Comparison between the mean recovery coefficients calculated for the predicted volumes obtained testing the CNN on TS1 (a), characterised by a PSF with FWHM = [4, 4, 4] mm, and on TS2 (b) characterised by a PSF with FWHM = [6, 6, 6] mm. A better recovery is obtained when PSFs are the same for training and test data.

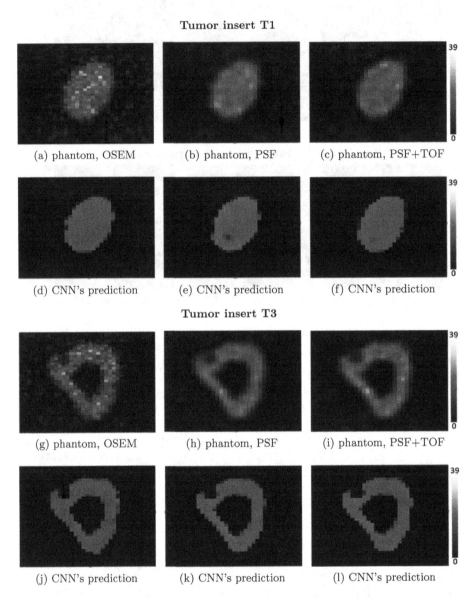

Tumor insert T1

(a) phantom, OSEM   (b) phantom, PSF   (c) phantom, PSF+TOF

(d) CNN's prediction   (e) CNN's prediction   (f) CNN's prediction

Tumor insert T3

(g) phantom, OSEM   (h) phantom, PSF   (i) phantom, PSF+TOF

(j) CNN's prediction   (k) CNN's prediction   (l) CNN's prediction

Fig. 6. The first and third row show three coronal views of the 3D patches extracted from phantom PET images, reconstructed using different algorithms (OSEM, PSF and PSF + TOF). Directly under each phantom image the corresponding CNN's prediction is presented. The intensities are expressed in kBq/mL.

Table 1. Maximum intensity values calculated for the real phantom scans and for the corresponding CNN's predicted images, expressed in kBq/mL.

	T1 phantom	T1 pred.	T2 phantom	T2 pred.	T3 phantom	T3 pred.
OSEM	34.77	19.19	31.79	18.18	38.78	21.72
PSF	26.65	20.90	24.15	23.98	27.19	22.30
PSF+TOF	25.34	21.76	25.35	20.89	27.56	21.08

4   Discussion and Conclusion

In this paper, we have developed a deep learning approach to improve tumor radiotracer quantification in PET images. A simulation algorithm was implemented to generate the labelled datasets needed for training. The algorithm can very effectively recover a more accurate estimate of the original distribution from the simulated PET images. It has been demonstrated that, for our purpose, the network was performing better when trained on non-normalized data. The predictions obtained by training the network on non-normalised data had better defined edges around the active volumes and the predicted intensities were more similar to the ground truth activity distributions. Simulated PET images generated using three different Gaussian blurring functions were included in the training set and, when tested on a dataset including the same three Gaussian functions, the CNN was able to correctly recover the ground truth images. This suggests that the network's performance does not have a strong dependence on the scanner-specific PSF. Preliminary results from applying the methods, trained on simulated data, to real phantom PET data are very encouraging. The main limitations of this method are the very simplified simulation of PET-like images and the poor knowledge of actual tumor ground truth distributions. The results presented in this work show that the proposed method has the potential to improve quantification of tumor tracer uptake, overcoming the challenges due to the lack of large labelled image datasets. This new quantification method could also be used as a part of an end-to-end image reconstruction process. In future work, we plan to train the network on more realistic images, simulated using more sophisticated simulation methods like Monte Carlo and to include some real patient data to the study.

References

1. Bai, B., Bading, J., Conti, P.S.: Tumor quantification in clinical positron emission tomography. Theranostics (2013). https://doi.org/10.7150/thno.5629
2. Baizán, A.N., Puig, D.R., Segura, J.P.: Evolution of quantification methods in oncologic 18F-FDG PET studies. Rev. Esp. Med. Nucl. Imagen Mol. **37**(4), 203–204 (2018). https://doi.org/10.1016/j.remn.2018.06.001
3. Blanc-Durand, P., Gucht, A., Schaefer, N., Itti, E., Prior, J.O.: Automatic lesion detection and segmentation of 18F-FET PET in gliomas: a full 3D U-Net convolutional neural network study. PLOS ONE **13**, e0195798 (2018). https://doi.org/10.1371/journal.pone.0195798

4. Gong, K., et al.: Iterative PET image reconstruction using convolutional neural network representation. IEEE Trans. Med. Imaging **38**(3), 675–685 (2019). https://doi.org/10.1109/TMI.2018.2869871

5. Gong, K., Guan, J., Liu, C.C., Qi, J.: PET image denoising using a deep neural network through fine tuning. IEEE Trans. Radiat. Plasma Med. Sci. **3**(2), 153–161 (2019). https://doi.org/10.1109/trpms.2018.2877644

6. Ioffe, S., Szegedy, C.: Batch normalization: accelerating deep network training by reducing internal covariate shift. In: Proceedings of the 32nd International Conference on International Conference on Machine Learning, ICML 2015, vol. 37, pp. 448–456. JMLR.org (2015). http://dl.acm.org/citation.cfm?id=3045118.3045167

7. Kim, K., et al.: Penalized PET reconstruction using deep learning prior and local linear fitting. IEEE Trans. Med. Imaging (2018). https://doi.org/10.1109/TMI.2018.2832613

8. Litjens, G., et al.: A survey on deep learning in medical image analysis. Med. Image Anal. (2017). https://doi.org/10.1016/j.media.2017.07.005

9. Loening, A., Sam Gambhir, S.: AMIDE: a free software tool for multimodality medical image analysis. Mol. Imaging **2**, 131–137 (2003). https://doi.org/10.1162/153535003322556877

10. Pfaehler, E., et al.: Repeatability of 18F-FDG PET radiomic features: a phantom study to explore sensitivity to image reconstruction settings, noise, and delineation method. Med. Phys. (2019). https://doi.org/10.1002/mp.13322

11. Strauss, L.G., Conti, P.S.: The applications of PET in clinical oncology. J. Nucl. Med. **32**, 623–648 (1991)

12. van der Vos, C.S., et al.: Quantification, improvement, and harmonization of small lesion detection with state-of-the-art PET. Eur. J. Nucl. Med. Mol. Imaging **44**, 4–16 (2017). https://doi.org/10.1007/s00259-017-3727-z

13. Wahl, R.L., Jacene, H., Kasamon, Y., Lodge, M.A.: From RECIST to PERCIST: evolving considerations for PET response criteria in solid tumors. J. Nucl. Med. **50**(Suppl. 1), 1–50 (2009). https://doi.org/10.2967/jnumed.108.057307

14. Wang, Z., Bovik, A.C., Sheikh, H.R., Simoncelli, E.P.: Image quality assessment: from error visibility to structural similarity. IEEE Trans. Image Process. **13**(4), 1–14 (2004). https://doi.org/10.1109/TIP.2003.819861

Task-GAN: Improving Generative Adversarial Network for Image Reconstruction

Jiahong Ouyang[1], Guanhua Wang[1,2], Enhao Gong[1,3], Kevin Chen[1],
John Pauly[1], and Greg Zaharchuk[1,3(✉)]

[1] Stanford University, Stanford, USA
gregz@stanford.edu
[2] Tsinghua University, Beijing, China
[3] Subtle Medical, Menlo Park, USA

Abstract. Generative Adversarial Network (GAN) has demonstrated great potentials in computer vision tasks such as image restoration. However, image restoration for specific scenarios, such as medical image enhancement is still facing challenge: How to ensure the visually plausible results while not containing hallucinated features that might jeopardize downstream tasks such as pathology identification? Here, we propose Task-GAN, a generalized model for medical reconstruction problem. A task-specific network that captures the diagnostic/pathology features, was added to couple the GAN based image reconstruction framework. Validated on multiple medical datasets, we demonstrated that the proposed method leads to improved deep learning based image reconstruction while preserving the detailed structure and diagnostic features.

1 Introduction

Image reconstruction in medical imaging is an important and attractive task since it enables imaging in more desirable conditions, e.g., imaging with faster protocols, cheaper devices, and lower radiation, etc. However, medical image restoration is still challenging as it requires not only visually realistic reconstruction, but also accurate image completion without altering pathological features or diagnostic qualities/properties.

Various related methods have been proposed recently, among which deep learning models, especially Generative Adversarial Network (GAN) [1] shows great potentials. Deep learning methods were used on Magnetic Resonance Imaging (MRI) reconstruction with aliasing inputs [2–4], low-dose Computer Tomography (CT) [5] and low-dose Positron Emission Tomography (PET) [6] reconstruction.

However, there are still several challenges and limitations for existing algorithms: (1) Pixel-wise losses do not consider non-local structural information thus leads to blurred and not visually plausible restoration. (2) GAN ensures the consistency to a learned distribution but do not necessarily guarantee the

© Springer Nature Switzerland AG 2019
F. Knoll et al. (Eds.): MLMIR 2019, LNCS 11905, pp. 193–204, 2019.
https://doi.org/10.1007/978-3-030-33843-5_18

visually plausible solution exactly matches the corresponding ground truth. (3) Discriminator network regularizes on general image distribution and visual quality, but it does not consider what are the characteristic features such as pathology and contrasts that the model needs to preserve.

Learning from works on computer vision tasks [7,8], adding an extra network that is specifically designed based on the property of the task should be helpful. Here, we proposed the Task-GAN, a generalized model for medical reconstruction problems. It includes 3 networks: a generator, a discriminator and a task-specific network. The new task-specific network predicts the pathology recognition from both the ground truth images and the reconstructed images. It helps to regularize the training of generator and complement the adversarial loss to ensure the output images better approximate the ground truth images. Task-GAN achieves realistic visual quality and preserves the important task-specific features/properties.

The contributions of this work are:

- We propose the Task-GAN method to ensure both visually plausible and more accurate medical image reconstruction.
- A task network and a task-driven loss are introduced to regularizes the image restoration to be more accurate both quantitatively and qualitatively.
- The proposed method was validated on two in-vivo clinical medical imaging datasets across different modalities, including MRI and PET.
- The theory behind the method is further discussed. The way of how the proposed Task-GAN improves the image reconstruction may lead to a better model design for other applications.

2   Proposed Method: Task-GAN

2.1   Designs

Here, we propose the Task-GAN that extends the GAN-based image reconstruction framework [9]. The goal of Task-GAN is to predict the image reconstruction of images X from the corrupted measurements $\tilde{X}$. In addition, we incorporate further information Y in the learning, which is one or a set of properties of X and important to preserve in the image restoration tasks. For example, the property can be the pathology or contrast specific feature in medical imaging, which are used as examples in this work. In general, there are three different networks that are optimized in the training process.

Firstly, a generator network G, learns the non-linear mapping from inputs $\tilde{X}$ to reconstructed images $\hat{X}$, which conducts the major image reconstruction task. This task is supervised with pixel-wise L_1 cost function, which has been shown outperform conventional L_2 cost function in image reconstruction tasks [9].

Secondly, a discriminator network D, similar to other adversarial training for GAN, is used to distinguish in the adversarial way to ensure $\hat{X}$ is consistent with the distribution of X. A classification task is conducted by D to learn $D(X) = 1$ and $D(\hat{X}) = D(G(X)) = 0$.

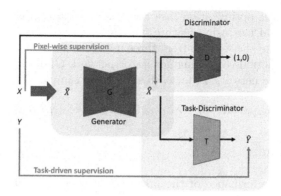

Fig. 1. Formulations and flowchart for Task-GAN

Lastly, a task-specific network T is generalized in the multiple image reconstruction settings. It tries to predict the set of properties of X, such that it favors $T(X) = Y$ and $T(\hat{X}) = Y$. For a binary case as in pathology recognition example in this work, $Y \in (0, 1)$, representing if there is pathology in the image. Other variants could include classifier for multi-label classification or segmentation network.

2.2 Formulation

Figure 1 shows the overall framework of the Task-GAN architecture for image restoration. For a single sample consisting of an image X with its property Y, we fed the corresponding corrupted image $\tilde{X}$ with the random noise z to the generator which outputs restored image $\hat{X}$. The weights of the three networks are optimized based on multiple cost functions across multiple tasks:

(1) To approximate the image content in X from generator G, we used pixel-level supervision with an L_1 loss between X and $\hat{X} = G(\tilde{X})$.

$$L_{pixel} = E_{(X,\tilde{X}) \sim p_{data}(X,\tilde{X}), z \sim p_z(z)} \left\| G(\tilde{X}, z) - X \right\|_1 \qquad (1)$$

(2) To stabilize the training process and to challenge the conventional discriminator Network D which recognizes whether the input is ground truth image X or restored version $\hat{X} = G(\tilde{X})$, we used the feature matching adversarial loss [10]. $f(X)$ denotes activations on an intermediate layer of the discriminator.

$$L_{GAN} = \left\| E_{X \sim p_{data}(X)} f(X) - E_{(\tilde{X}) \sim p_{data}(\tilde{X}), z \sim p_z(z)} f(G(\tilde{X}, z)) \right\|_2 \qquad (2)$$

(3) To teach the Task network T to recognize the property Y from image X, and more importantly to ensure that the recognizable features are still preserved in $\hat{X}$, we use a regression loss for this task.

$$L_{task} = E_{(X,\tilde{X},Y) \sim p_{data}(X,\tilde{X},Y), z \sim p_z(z)} (T(G(\tilde{X}, z)) - Y)^2 + (T(X) - Y)^2 \qquad (3)$$

In summary, the optimization task consists of 3 parts: pixel-wise loss using L_1 cost, adversarial loss using feature matching GAN cost and task loss with regression cost. And the weights from three networks are optimized to minimize the mixed loss function combining generator network G, discriminator network D and task-specific network T, for each supervised sample $(X, \tilde{X}, Y)$

$$L(G, D, T) = L_{pixel}(G, X, \tilde{X}) + \lambda L_{GAN}(G, D, X, \tilde{X}) + \mu L_{task}(T, X, \tilde{X}, Y) \quad (4)$$

3   Experiments

To evaluate the performance of the proposed method, we did experiments on two tasks: ultra-low-dose amyloid PET reconstruction and multi-contrast MR reconstruction. Here, we briefly describe the purpose of the tasks, the design of task-specific networks T, the datasets, and the comparison results.

3.1   Ultra-low-dose Amyloid PET Reconstruction Task

Task & Task-Specific Network: The purpose of this task is to synthesize high-quality standard-dose amyloid PET from 1% ultra-low-dose degraded PET images, preserving pathological features Y of the amyloid status (positive or negative) that related to the diagnosis of Alzheimer's Disease. Thus, in this task, we pre-trained an 2D amyloid status classifier on the standard-dose PET images as T (based on standard-dose image) to extract pathological features Y and optimized the perceptual loss [11]. Two experts' amyloid status diagnosis were used as ground-truth label for the classifier. The state-of-the-art method using L_1 loss proposed by Chen et al. [6] was implemented as reference.

Data Acquisition & Prepossessing: 40 subjects were recruited for the study, among which 10 subjects were amyloid status positive and the other 30 were negative. Datasets were acquired on an integrated PET/MR scanner with time-of-flight capabilities (SIGNA PET/MR, GE healthcare). 330 ± 30 MBq of the amyloid radiotracer (18F-florbetaben) was injected to the subject (as standard-dose) and the PET data was acquired 90–110 min after injection. The raw list-mode PET data was reconstructed as the standard-dose PET and it was randomly undersampled by a factor of 100 to reconstructed the low-dose PET. Each PET volume consists of 89 2.78 mm-thick slices with 256×256 1.17×1.17 mm^2 pixels. Each volume was normalized by the mean value of the non-zero region. The top and bottom 20 slices, which usually did not cover the brain, were removed. To avoid overfitting, dara augmentation of flipping along the X and Y axes was adopted. Four-fold validation was adopted to obtain synthesized results for each data.

Results: A qualitative comparison of the reconstructed image is shown in Fig. 2. Radiologists are trained to make diagnosis of amyloid status positive/negative from the detailed tracer retention pattern in cortex area. Figure 2-A is amyloid negative and B is amyloid positive. Chen et al.'s method and Plain-GAN blurred out some parts of cortex and generated some hallucinate uptake, while the proposed method kept the original pathological structures better. This is especially significant in amyloid status negative cases like, as radiologists intend to diagnose them as positive due to the blurriness.

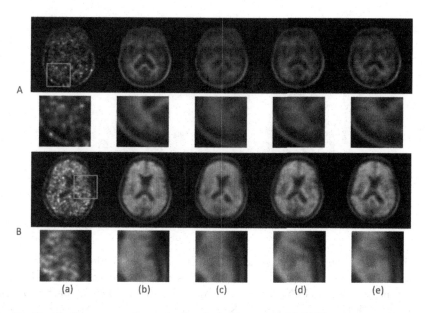

Fig. 2. Qualitative comparison of ultra low-dose amyloid PET reconstruction task: (a) low-dose PET, (b) standard-dose PET, (c) Chen et al., (d) Plain-GAN, and (e) Task-GAN.

The quantitative results on image quality are shown in Fig. 3. We evaluated the synthesized image quality by peak signal-to-noise ratio (PSNR), structural similarity (SSIM), and root mean square error (RMSE). The proposed method significantly outperforms the plain-GAN without the task network by 1.62 dB in PSNR, 1.58% in SSIM, and 16.4% in RMSE. We further evaluated the image quality by reader study, which is shown in Fig. 4. Two experts was asked to give a score 1–5 as the quality score for each volume (total 80 evaluations of 40 cases by 2 readers). We considered 1–3 as low quality and 4–5 as high quality. The results from proposed method had an average score of 4.27 with only 5 low quality, which was comparable to the ground-truth 4.41 with 4 low quality, and significantly outperformed Chen et al.'s 3.22 with 56 low quality scores.

Moreover, we evaluated how the proposed method helped down-stream task: amyloid status diagnosis. Table 1 shows the classification results given by the 2D amyloid status classifier T. Based on the classification on the middle 20 slices of the volume, the subject-wise error rate is computed by voting and following the

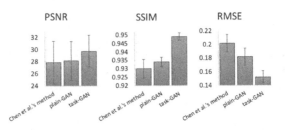

Fig. 3. Image quality metrics of low-dose PET reconstruction task.

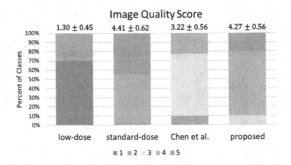

Fig. 4. Image quality score of low-dose PET reconstruction task.

majority rule. Though the classifier achieved all correct subject-wise diagnosis, the slice-wise mean average error (MAE) illustrated more accurate and stable performance on Task-GAN outputs. Radiologists also achieved better performance on Task-GAN's results with 8 errors over 80 diagnosis, comparing to 16 errors on Chen et al.'s results.

Table 1. Diagnosis by amyloid status classifier T.

	Standard-dose PET	Chen et al.	Plain-GAN	Task-GAN
slice-wise MAE	0.136 ± 0.121	0.140 ± 0.138	0.138 ± 0.132	0.132 ± 0.118
subject-wise error	(0/40)	(0/40)	(0/40)	(0/40)

3.2   Multi-contrast MR Reconstruction Task

Task & Task-Specific Network: To validate the generality of the proposed method on different modalities of medical imaging restoration, we conducted experiments on synthesized multi-contrast magnetic resonance imaging (MRI) datasets. The real measurements were acquired at different Echo time (TE), Inversion time (TI) and Delay time (TR). Traditionally, the model-based method uses least-square fitting to resolve T1, T2, PD and B1 map, and then synthesize different clinical contrasts (T1w, T2w, FLAIR, etc.) retrospectively. In this setting, we tried to reconstruct high-quality multi-contrast MR neuroimaging from the multi-dynamic multi-echo acquisition (MDME) inputs [13] while preserving the contrast-specific features. The proposed method is aimed to not only generate visually plausible MR images but also preserve the parameter-weighted contrast that required for the diagnosis of neurological diseases. The reconstruction of multiple contrasts MR shares the weight and the specific contrast can be synthesized by changing the binary-coded vector that added to the model. The task-specific network T in this setting is a classifier that can discriminate different contrasts, assisting the generator in learning the features of different contrasts. At the same time there is also a binary discriminator as in the traditional GAN, sharing the same feature extraction layers.

Data Acquisition & Preprocessing: 109 cases were included in the datasets. Among them, 61 were patients and 48 were healthy controls. Each subject was scanned with 6 conventional MR sequences(T1w, T1-FLAIR, T2w, T2-FLAIR, STIR and PDw) as the ground-truth and MDME sequences as the generator's input. Traditional model-based method [14] was also calculated as reference. Protocols include FOV: 230 * 184 mm, thickness: 5 mm, number of slices: 25. During the pre-processing, different acquisitions were affinely registered. The magnitude of image was re-scaled to [0, 1]. Real and imaginary part were formulated as two input channels.

Results: The qualitative comparison of a typical examples are shown in Fig. 5, in which plain-GAN wrongly learned and introduced the artificial features like the grid-like artifacts, while the proposed model achieved better detail-restoration and sharpness. Using Task-GAN, the generator can get information from contrasts whose quality is better, and store the shared information in the encoded latent space, to correct imperfection like the motion and partial volume effect in contrasts like FLAIR.

The quantitative comparison of similarity with conventional acquisitions is shown in Fig. 6, the proposed method improved PSNR by 3.95 dB, SSIM by 9.1% compared to the traditional model-based method. In comparison with plain-GAN, the Task-GAN raised PSNR by 2.40 dB, SSIM by 7.7%. In addition, 4 radiologists were asked to rate the image quality with Likert score from 1 ('unacceptable') to 5 ('excellent'). The paired Student's t-test was performed to compared the scoring. The results are shown in Fig. 7.

4   Discussion

The proposed method achieves superior performance on in-vivo medical imaging datasets by coupling adversarial training with the task-specific network. Detailed contribution of the Task-GAN is explained in Fig. 8.

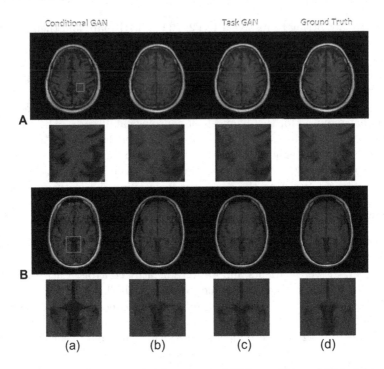

Fig. 5. Qualitative comparison of Multi-contrast MRI experiment: (a) Model-based (b) plain-GAN (c) Task-GAN, (d) conventional acquisitions (ground-truth).

In comparison, the task of the image reconstruction is to learn a non-linear mapping from low-quality images in the measurement domain to its corresponding high-quality images in a different high-quality domain containing visually realistic images. Shown in Fig. 8(a), in addition, the recognition of image is a space separation of features/labels along different dimensions that can be orthogonal to the quality dimensions.

In comparison, as is shown in Fig. 8(b), conventional learning strategy learns the image reconstruction task by regression, which may fail to generate realistic reconstruction. The learning is usually based on the minimization of an averaged

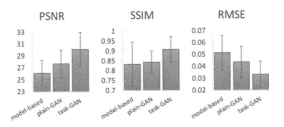

Fig. 6. Image quality metrics of multi-contrast MR reconstruction task.

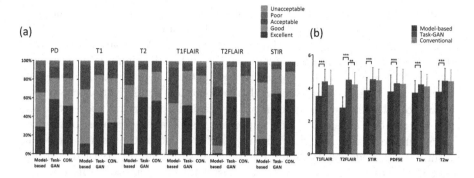

Fig. 7. Image quality score of multi-contrast MR reconstruction task. (a) detailed the proportion of different scores. (b) depicts the mean score. ** means the p value is under 0.005, while *** stands for 0.0005.

distance penalty which ensures robustness but lead to unrealistic reconstruction such as blurring. This can be seen from Fig. 2 Chen et al.'s results. Additionally, the averaged solution is also likely to be away from the distribution of visually plausible solutions that falls out of the high-quality image space as is shown in the Fig. 8(b).

GAN-based approach on one hand overcomes this by further enforcing an adversarial loss with a discriminator network which ensures to generate realistic reconstruction following the distribution of the target high-quality images. As the Fig. 8(c) shows, the solution is no longer an simple average, but pushed into the space of visually realistic high-quality images. However, on the other hand, the discriminator only regularizes the output samples to follow the distribution while ignores the inter-sample relationship. For example, it cannot avoid hallucinations, where the restored images may be undesirably added/removed important visual features. As is shown in Fig. 8(c), the output image can have a different label as the ground-truth which fails the purpose of image reconstruction. We can picture the hallucinations as a "shrinking" of solution space.

To ensure the one-to-one mapping, various improved GAN models and cost functions have been proposed. For example, Cycle-GAN [12] incorporate a cyclic relationship to improve the mapping. However, cyclic relationship does not necessarily lead to exact mappings. As Fig. 8(d) demonstrates, the inter-sample relationship as well as the important feature labels can be swapped while still satisfying the cyclic relationship. For example, in the figure, one task label is altered while the cyclic loss is not affected. This may lead to misclassified pathology/normality for medical imaging applications, which can directly lead to misdiagnosis or over-diagnosis. We can picture the mislabeling as a "twisting" of solution space. This "twisting" maintains well within visually-plausible space, however severely changes the positioning around the decision boundary of task-label space.

Differently, as shown in Fig. 8(e), Task-GAN can generate accurate mapping with the mixed loss regularization: (1) Pixel-level supervision ensures that the reconstructed image is closer to the ground truth. (2) Adversarial loss regularization ensures that the reconstructed image is within the high-quality space. (3) Task-specific loss ensures the reconstructed image still preserve the important feature of interests. In other words, the combination regularization enforce the solution to fall onto the intersection of the manifold preserving pixel-level similarity, distribution consistency and important visual labels/features. In the view of inter-sample relationship, the task regularization stops the inter-sample relationship to any visual plausible but destructive "shrinking" or "twisting" around the boundary of task-label space, which ensures more accurate mappings.

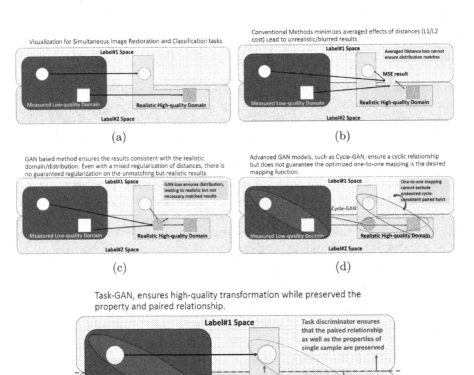

Fig. 8. How Task-GAN improves the mapping in image restoration.

5   Conclusion

In this paper, we proposed a generalized model, Task-GAN, which includes a new task-specific network and corresponding task-specific loss for training GAN based image reconstruction. Task-GAN is demonstrated to boost the performance of medical image reconstruction, which requires not only realistic restoration, but also high-fidelity as well as accurate classification for subtle diagnostic features.

The proposed method is demonstrated to achieve superior performance compared with GAN on both image quality metrics and task-specific feature preservation (e.g. pathological features). Based on visual inspection from human experts (clinicians/radiologists), anatomical and diagnostic features are preserved better and fewer artifacts are introduced. The trained task network also shows potentials for super-human level diagnosis tasks.

Task-GAN further extends the regularization of adversarial training. The mixed loss balances between content similarity, distribution consistency and preserving important features for the given tasks. It results in more accurate image reconstruction with better visual similarity and avoids hallucinations. Intuitively, task-GAN enforces the solution fall into proper manifold, prevents any alternation ("shrinking" and "twisting") of the reconstruction from the correct solution space, and preserves both inter-sample relationship and feature-of-interest.

In the future, we will explore further improvements in the design of networks and task formulation. The proposed technique is also valuable to other challenging reconstruction/restoration applications that require realistic restoration and preserving distinguishable details for down-stream tasks.

References

1. Goodfellow, I., et al.: Generative adversarial nets. In: NIPS (2014)
2. Mardani, M., et al.: Deep generative adversarial neural networks for compressive sensing MRI. IEEE Trans. Med. Imaging **38**(1), 167–179 (2019)
3. Seitzer, M., et al.: Adversarial and perceptual refinement for compressed sensing MRI reconstruction. In: Frangi, A.F., Schnabel, J.A., Davatzikos, C., Alberola-López, C., Fichtinger, G. (eds.) MICCAI 2018. LNCS, vol. 11070, pp. 232–240. Springer, Cham (2018). https://doi.org/10.1007/978-3-030-00928-1_27
4. Yang, G., et al.: DAGAN: deep de-aliasing generative adversarial networks for fast compressed sensing MRI reconstruction. IEEE Trans. Med. Imaging **37**(6), 1310–1321 (2017)
5. Kang, E., et al.: A deep convolutional neural network using directional wavelets for low-dose X-ray CT reconstruction. Med. Phys. **44**(10), e360–e375 (2017)
6. Chen, K., et al.: Ultra-low-dose 18F-florbetaben amyloid PET imaging using deep learning with multi-contrast MRI inputs. Radiology. **290**(3), 649–656 (2018)
7. Huang, R., et al.: Beyond face rotation: global and local perception GAN for photorealistic and identity preserving frontal view synthesis. In: ICCV (2017)
8. Hoffman, J., et al.: CyCADA: cycle-consistent adversarial domain adaptation. In: ICML (2018)
9. Isola, P., et al.: Image-to-image translation with conditional adversarial networks. In: NIPS (2016)

10. Salimans, T., et al.: Improved techniques for training GANs. In: NIPS (2016)
11. Johnson, J., Alahi, A., Fei-Fei, L.: Perceptual losses for real-time style transfer and super-resolution. In: Leibe, B., Matas, J., Sebe, N., Welling, M. (eds.) ECCV 2016. LNCS, vol. 9906, pp. 694–711. Springer, Cham (2016). https://doi.org/10.1007/978-3-319-46475-6_43
12. Zhu, J., et al.: Unpaired image-to-image translation using cycle-consistent adversarial networks. arXiv preprint arXiv:1703.10593 (2017)
13. Warntjes, J.B.M., et al.: Rapid magnetic resonance quantification on the brain: optimization for clinical usage. Magn. Reson. Med. 60, 320–329 (2008)
14. Tanenbaum, L.N., et al.: Synthetic MRI for clinical neuroimaging: results of the Magnetic Resonance Image Compilation (MAGiC) prospective, multicenter, multireader trial. Am. J. Neuroradiol. 38, 1103–1110 (2017)

Gamma Source Location Learning from Synthetic Multi-pinhole Collimator Data

Peter A. von Niederhäusern[1]([⊠]), Carlo Seppi[1], Simon Pezold[1],
Guillaume Nicolas[2], Spyridon Gkoumas[3], Stephan K. Haerle[4],
and Philippe C. Cattin[1]

[1] Department of Biomedical Engineering, University of Basel, Allschwil, Switzerland
`peter.vonniederhaeusern@unibas.ch`
[2] University Hospital of Basel, Radiology and Nuclear Medicine Clinic,
Basel, Switzerland
[3] DECTRIS Ltd., Baden-Dättwil, Switzerland
[4] Center for Head and Neck Surgical Oncology and Reconstructive Surgery,
Hirslanden Clinic, Lucerne, Switzerland

Abstract. Sentinel lymph node biopsy (SNB) is a surgical method to
stage certain cancer types in a minimally invasive manner. However, the
current sensing methods for SNB are limited in accuracy, as they are
based on acoustic feedback radiation probes to detect tracer enriched
sentinel lymph nodes. We present a deep neural network approach to
learn the latent spatial activity distributions from a simulated gamma
source on 2D activity images. Data processing can then be applied for
multi-pinhole collimator optimization, lymph node visualization or surgical navigation to further support SNB. Using simulations of photon
multi-pinhole collimator interaction, we generate labeled synthetic 2D
activity images to train convolutional neural networks (CNN). These
CNNs are then evaluated on synthetic as well as on real experimental
data from a radioactive point-like source, collected by our own stationary
small form factor multi-pinhole collimator. We achieve good results on
synthetic data for the xy-component ensemble learners with a localization class accuracy of 0.97, while depth estimation achieves a localization
class accuracy of 0.55. Accuracy on real experimental data is limited due
to the small sample set and its variability, compared to the simulation.

Keywords: Sentinel lymph node biopsy · Radioguided surgery ·
Inverse problem · Machine learning · Convolutional neural network

1 Introduction

In head and neck squamous cell carcinoma (HNSCC) the standard treatment
comprises the elimination of the tumor together with the nearby lymphatics to
ensure the removal of malignant cells to prevent further spreading. This procedure is called neck dissection and is highly invasive where overtreatment is
tolerated. However, studies have shown that such radical interventions are not

© Springer Nature Switzerland AG 2019
F. Knoll et al. (Eds.): MLMIR 2019, LNCS 11905, pp. 205–214, 2019.
https://doi.org/10.1007/978-3-030-33843-5_19

needed in 70% of patients [1]. Sentinel lymph node biopsy (SNB) is a surgical procedure to stage the malignancy in a minimally invasive manner to avoid this overtreatment. The current sensing methods for SNB are still rather limited in accuracy, as acoustic feedback gamma radiation probes are used to detect radioactive tracer enriched sentinel lymph nodes (SLN) while conducting the actual biopsy. The size of a lymph node is typically between 5–10 mm and therefore easy to miss. This puts a high cognitive load on the surgeon as they need to map the audio signal (1-dimensional) from the probe to the tissue surface of the patient to successfully place the biopsy tools. Freehand SPECT (fhSPECT) is an alternative approach to reconstruct and visualize activity [2]. One drawback of fhSPECT is the need to manually rescan the surgical scene from different viewpoints to update the visualization in case of tissue removal or patient repositioning. Our vision is a setup with a stationary multi-pinhole collimator/gamma camera attached near the patient in order to acquire activity and to render its distribution subsequently. We therefore circumvent the inconvenience of manual scanning during intervention. Current research in the domain to reconstruct spatial activity distributions from 2D activity images tries to solve this inherently ill-posed problem by the application of compressed sensing [3] or fingerprinting [4]. These methods require a significant amount of RAM and compute to store and process the system matrix: compressed sensing is rather slow and memory-intensive while fingerprinting is fast but also memory-demanding. If the measurement space increases, these requirements get accentuated.

In this work, we present a simulation and a recovery method to learn the spatial components of a γ-source, e.g. a lymph node, from 2D activity images alone, i.e. without using tomographic information. These images are produced by our stationary custom-built multi-pinhole collimator (Fig. 6(a)), in combination with a compact gamma camera (Sect. 2). Fundamental advantages of a multi-pinhole arrangement are its capability to allow for depth estimation and increased photon sensitivity in comparison to a standard single pinhole collimator: photons from different viewpoints are collected and contribute to the foreground signal. The spatial learning itself is achieved by three independent CNNs (Sect. 2). To obtain training data, we developed an algorithm to simulate interaction between incident photons of a synthetic γ-source with the modeled multi-pinhole collimator (Sect. 2). We use both simulated and real 2D activity images to evaluate the performance of the trained networks (Sect. 3).

2   Methods and Materials

Problem Formulation. A forward problem is often of the form

$$A(v) \Rightarrow I. \tag{1}$$

The modeling operator A corresponds in our problem to the geometry of the multi-pinhole collimator. The parameters v represent the spatial distribution of the source, and the observation I is the actual 2D activity image. As we normally do not know v, spatial information recovery is only possible using I, which is an ill-posed problem. Regularization is needed to gain additional information and constraints. In particular, the assumption is that there exists a sparse solution to simplify the inverse problem. We formulate the reconstruction of a potentially unknown spatial distribution of γ-activity from a 2D activity image I by

$$\hat{A}(I) \Rightarrow \hat{v}, \tag{2}$$

where $\hat{A}$ is the pseudo-inverse of A, and $\hat{v}$ is an approximation of the true source location. In our case, spatial estimation is done by applying three independent CNNs to learn $\hat{A}$ from I:

$$\hat{v}_i = \text{CNN}_i(I), \ i = x, y, z. \tag{3}$$

Regularization is implicitly achieved by the network design and training process of these CNNs.

Learning. We define γ-source location learning as a classification problem. The covered depth $[80, 180]$ mm is mapped to 21 classes resulting in an uncertainty of 5 mm within each class c_z. We proceed similarly for the xy-components. The x-component is given 21 classes, covering a physical space $[-30, 30]$ mm with an uncertainty of 3 mm. The range for the y-component $[-21, 21]$ mm is partitioned into 15 classes with an uncertainty of 3 mm. The setup of the measurement space is shown in Fig. 4. To consider the influence of the span per class bin, we apply five shifts of 1 mm for c_{z_j} and three shifts of 1 mm for both c_{x_j} and c_{y_j}. Each shifted class bin has therefore a slightly different mean value. As such, multiple mean values of the ensemble contribute to an estimate of the spatial components. We train the depth ensemble learner CNN_z five times, and the xy-component ensemble learners $\text{CNN}_{x|y}$ three times with these different class mappings.

As an example, we take an instance of c_z with a known distance of 95 mm. The correctly predicted class instances $\hat{c}_{z_j}$ have bin ranges $[91, 96), [92, 97), [93, 98),$ $[94, 99), [95, 100)$ mm with rounded averages $\in [93.5, 94.5, 95.5, 96.5, 97.5]$ mm. The estimated depth value is thus $\bar{\hat{c}}_z = \lfloor 95.5 \rfloor$ mm. In this manner we proceed to estimate $\hat{c}_{x_j}$ and $\hat{c}_{y_j}$. These estimates are then compared to the known ground truth location (Sect. 3).

Nearest Neighbor Lookup (NNL). Given training data, a k-nearest neighbor search can be used to find similar synthetic and real images. We set $k = 1$. The metric function applied is the scalar product: a minimal angle indicates matching features. The classified xyz-components of the lookup are compared to real experimental data

$$\hat{v} = \text{NNL}_{k=1}(I). \tag{4}$$

This brute force algorithm serves as a baseline for our CNN-based approach.

Sparse Signal Recovery. The authors of [3] provided us with their implementation of the WSPGL1 (weighted spectral projected gradient for ℓ_1 minimization) algorithm for sparse signal recovery, as described in [5]. We compare results on real experimental data.

Image Formation. The field of view of a single pinhole is constrained by the radius of the pinhole itself and the compartment dimensions (Fig. 1). The half-angle α of the field of view is given by

$$\tan(\alpha) = \frac{w}{2} \cdot \frac{1}{h},$$ (5)

where w denotes the width and h the height of the pinhole compartment. Based on our design, the field of view of a pinhole is $2 \cdot \alpha \approx 16°$.

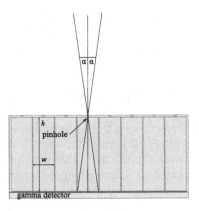

Fig. 1. Collimator compartment arrangement with dimensions height h (length of the sidewalls or septa) and width w. The field of view $2 \cdot \alpha$ of the pinholes depends on these measures. Photons are incident from the top, the gamma camera sensor is at the bottom (indicated in red). (Color figure online)

Typical real photon-sensor interactions of two distinct point source positions (90 mm, 150 mm) are presented in Figs. 2(a) and 3(a). Each pinhole with a suited field of view collimates photons and projects them onto the image sensor. Unwanted photon collimator interactions are mainly caused by Compton and photoelectric effects. In order to constrain high-energy Compton scattering, we rely on the compartment length or cross section of the collimator for absorption (*cf.* Sect. Hardware). To compensate for image degradation by the low-energy photoelectric effect, we apply filtering (*cf.* Sect. Image preprocessing). Similar simulated activities are shown in Figs. 2(b) and 3(b).

One compartment can be seen as a region having a distinct view of the source. A source farther away is captured by more pinholes, while its projections move inwards to the center of the compartment. This disparity effect can be exploited and statistically analyzed to learn depth information.

Data Generation. As the real experimental data set for realistic training is too small, we developed a model simulation of photon collimator interaction.

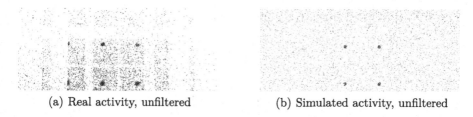

(a) Real activity, unfiltered   (b) Simulated activity, unfiltered

Fig. 2. Activity image of a single point source, centered at a distance of 90 mm. Each pinhole projects captured photons from the source onto the image plane. Red pixels indicate high photon counts (≥ 6). (a) Real source, 60 MBq, acquisition time 16 s; due to a slight misalignment of the collimator relative to the sensor, partial pinhole projections (leftmost column) are discernible. The dark patches are caused by high energy photons penetrating the front shield near the pinholes, and show the absorbing effect of the compartment septa. (b) Simulation with added background signal. (Color figure online)

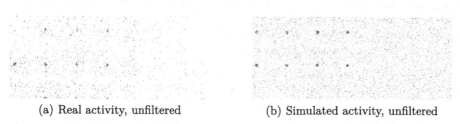

(a) Real activity, unfiltered   (b) Simulated activity, unfiltered

Fig. 3. Activity image of a single point source, positioned left at a distance of 150 mm. (a) Real source, 60 MBq, acquisition time 16 s. The detector has one faulty ASIC element in the upper right corner. (b) Simulation with added background signal.

Evaluating whether a simulated γ-ray reaches a pixel of the virtual image plane, under some constraints explained below, leads to a good approximation of the actual activity data (*cf.* Figs. 2(b) and 3(b)). We refrain from using Monte Carlo simulation due to the slow and expensive computation involved.

From the line-plane intersection test we know whether a simulated γ-ray intersects the aperture mask, and the intersection point can thus be calculated. However, we still need to determine whether the γ-ray passes through a pinhole. As the coordinates of the pinholes are known, this test is trivial. A last constraint which must be considered is whether the incident angle of the γ-ray is steeper than the half angle α of the field of view; the γ-ray is either collimated to reach the sensor or absorbed by the septum (*cf.* Fig. 1). We first define the following:

- Let $\mathbf{p}_i$ be a pixel on the virtual image plane.
- Let $\mathbf{s}_j$ be a discrete element of the synthetic source $\mathbf{S}$.
- Let $\mathbf{i}$ be the intersection point on the aperture mask.
- Let h be the height of the compartment.

We calculate the angles β_x, β_y of the incident γ-ray

$$\beta_x = \tan\left(\frac{1}{h}\,|\mathbf{p}_{i_x} - \mathbf{i}_x|\right), \tag{6}$$

$$\beta_y = \tan\left(\frac{1}{h}\,|\mathbf{p}_{i_y} - \mathbf{i}_y|\right) \tag{7}$$

and update the pixel on the virtual image plane

$$\text{increment}(\mathbf{p}_i) = \begin{cases} 1, & \text{if } \beta_x \text{ and } \beta_y \leq \alpha, \\ 0, & \text{otherwise.} \end{cases} \tag{8}$$

We do the above for every combination of $\mathbf{s}_j$ and $\mathbf{p}_i$. Each simulation consists of one source $\mathbf{S}$ with varying radius $\in [0.5, 3]$ mm, xy-position and distance (*cf.* Sect. Learning). We added some randomness to the process to allow a 5% probability for simulated photons to interact with the collimator material and to contribute to the background signal (*cf.* Sect. Hardware). The result is a synthetic 2D activity image $I(\mathbf{p})$ which serves as a training sample.

Experimental Data. Real activity data was acquired using one technetium (^{99m}Tc) tracer source of 60 MBq with an acquisition time of 16 s. To direct the radiation in a point-like fashion, the source container was fixed in a Cerrobend block with exit pupils of radius 1 mm, pointed at the multi-pinhole collimator. We measured four distances (90 mm, 110 mm, 130 mm, 150 mm) with three varying horizontal and vertical offsets relative to the collimator coordinate system (Fig. 4). Note that the real experimental data set is limited to 12 samples (*cf.* Sect. 3).

Image Preprocessing. For every 2D activity image, we apply a median filter of size 3×3 to reduce the influence of the background signal. A Gaussian filter of size 3×3 is used to increase signal response. We consider the full image extent of 487×195 pixels to yield the input vector to the $\text{CNN}_{x|y}$ for xy-component learning. This differs for depth estimation: each image region of a compartment (*cf.* Fig. 5(a)) is extracted and stacked to form a volume of activities (*cf.* Fig. 5(b)). Finally, we sum up the volume slices to create a unique map of the depth activity distribution (*cf.* Fig. 5(c)). High signals or sums, distributed over the patch area of the map, indicate an accumulation of projections. The CNN_z then learns based on these projections.

Network Design. The CNN architectures are given below, where C indicates a convolutional layer with a prefix for the number of filter banks and a suffix for the filter size, BN denotes a batch normalization step, MP is a max. pooling operation, DO a transition with a given dropout ratio, FL a matrix flattening, and N a dense layer for the final classification.

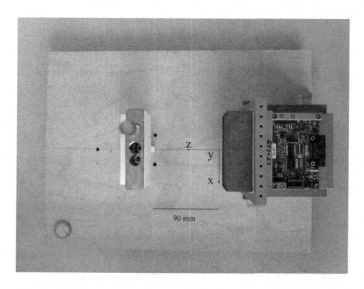

Fig. 4. Experimental setup, viewed from top. The Cerrobend block with the inserted radioactive source is on the left (red vial). The stationary detector with the attached collimator is on the right. Here, the known distance from the collimator to the source is 90 mm. The collimator measurement space, with its coordinate system at the origin, is given. A 1-€ coin in the lower left serves as a scale reference. (Color figure online)

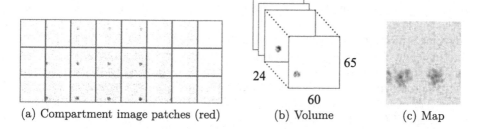

(a) Compartment image patches (red)   (b) Volume   (c) Map

Fig. 5. (a) Each compartment image patch forms a slice of the volume. (b) Volume of size $65 \times 60 \times 24$. (c) Sum of the volume slices (projection along the depth) to create a map of size $65 \times 60 \times 1$, fed into the CNN_z. (Color figure online)

$$65 \times 60 \times 1 \rightarrow$$
$$16\text{C}5 \rightarrow \text{BN} \rightarrow \text{ReLU} \rightarrow \text{MP}2 \rightarrow \text{DO}(0.5) \rightarrow$$
$$32\text{C}5 \rightarrow \text{BN} \rightarrow \text{ReLU} \rightarrow \text{MP}2 \rightarrow \text{DO}(0.5) \rightarrow$$
$$64\text{C}3 \rightarrow \text{BN} \rightarrow \text{ReLU} \rightarrow \text{MP}2 \rightarrow \text{DO}(0.5) \rightarrow$$
$$128\text{C}3 \rightarrow \text{BN} \rightarrow \text{ReLU} \rightarrow \text{MP}2 \rightarrow \text{DO}(0.5) \rightarrow$$
$$\text{FL} \rightarrow 512\text{N} \rightarrow \text{BN} \rightarrow \text{ReLU} \rightarrow \text{DO}(0.5) \rightarrow$$
$$21\text{N} \rightarrow \text{Softmax}$$

Listing 1. The network architecture of CNN_z for depth learning.

$$487 \times 195 \times 1 \rightarrow$$
$$16C5 \rightarrow BN \rightarrow ReLU \rightarrow MP2 \rightarrow DO(0.5) \rightarrow$$
$$32C5 \rightarrow BN \rightarrow ReLU \rightarrow MP2 \rightarrow DO(0.5) \rightarrow$$
$$64C3 \rightarrow BN \rightarrow ReLU \rightarrow MP2 \rightarrow DO(0.5) \rightarrow$$
$$128C3 \rightarrow BN \rightarrow ReLU \rightarrow MP2 \rightarrow DO(0.5) \rightarrow$$
$$FL \rightarrow 512N \rightarrow BN \rightarrow ReLU \rightarrow DO(0.5) \rightarrow$$
$$(21_x \mid 15_y)N \rightarrow Softmax$$

Listing 2. The network architecture of $CNN_{x|y}$ to learn the xy-component of the activity.

The training set consists of 43'000 synthetic 2D activity images with an 80/20 training/validation ratio. The test set contains 6'000 samples. We trained the network for 100 epochs with a batch size of 32. As optimizer we chose Adam with learning rates $\in$ [1e–03, 1e–07]. The learning rate was gradually reduced based on the validation accuracy during training.

Hardware. The current design of the actual multi-pinhole collimator is shown in Fig. 6. The frame is a 3D-printed tungsten alloy, selective laser melted, with a mass attenuation coefficient of $\approx$1.581 at 140.0 keV. The radius of each cylindrical pinhole is 0.5 mm, and its associated compartment with width 10.0 mm, length 35.0 mm defining the field of view of $\approx$16° (*cf.* Fig. 1). The thickness of the front plate of 1.0 mm and the thickness of the septa of 0.35 mm with length 35.0 mm are calculated such, that the probability of background photons to penetrate the shielding is at most 5%. The dimensions of the device are 83.94 mm in width, 33.54 mm in height and 35.0 mm in depth, respectively. Its weight is 300 g. Our industrial partner DECTRIS (Baden-Dättwil, Switzerland) provided us with a gamma camera prototype with a native resolution of 487 × 195 pixels and a pixel size of 172 μm × 172 μm. The detector technology of DECTRIS is based on Hybrid Photon Counting (HPC) with a cadmium telluride (CdTe) sensor material [6]. Its quantum efficiency (QE) at 140.0 keV is $\approx$32%. A high QE is crucial as only 1% or less of the injected tracer activity arrives in the lymphatics.

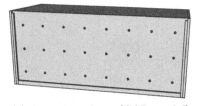   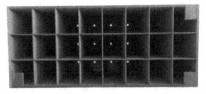

(a) Collimator front (CAD model)   (b) Collimator back (real, 3D-printed)

Fig. 6. View of the collimator. (a) Aperture mask with its pinhole arrangement. (b) Compartments with calibration plates in the corners. The pinhole collimating effect is observed by the passing light in the central part.

3   Results

For all results, accuracy indicates a perfect class match; only the exact classification, *i.e.* bin, is considered. Near class misses are therefore equally weighted as estimates with a larger class mismatch. We show the *test set* class accuracy of each ensemble $CNN_{x|y|z}$ in Table 1.

Table 1. Individual and average class accuracy of the learners on the test set.

Learner	CNN_x	CNN_y	CNN_z
# 1	0.97	0.95	0.61
# 2	0.98	0.98	0.46
# 3	0.97	0.99	0.62
# 4	–	–	0.48
# 5	–	–	0.58
Average	0.97	0.97	0.55

We achieve a high score for CNN_x and CNN_y, with learners 1–3. Depth learning, with learners 1–5, scores lower, as it is a harder problem. This can also be seen in Table 2, where depth estimation is the most error-prone task for spatial component analysis.

In Table 2 we present the overall results from the *real experimental data set*. We win against both WSPGL1 and NNL in terms of the number of minimal error scores achieved: 5 out of 12. Our GPU-accelerated method is the fastest with an inference time of CNN $\approx$1.0 s/sample, compared to WSPGL1 $\approx$100.0 s/sample and NNL $\approx$60.0 s/sample, which both run on the CPU.

Table 2. For each sample from the real experimental data set, we compare the known location with the inferred location, per method. Location components are (x, y, z), given in mm. The errors are reported using the ℓ_2-norm between the true location and the inferred location. The overall minimal error is marked in bold.

Methods				Error: ℓ_2		
Location	NNL	WSPGL1	CNN	NNL	WSPGL1	CNN
(0, 5, 90)	$(-2, 4, 95)$	$(-0.5, 11.5, 88)$	$(-1.5, 4.5, 89.3)$	5.5	6.8	**1.7**
(0, 5, 110)	$(-2, 5, 97)$	$(-0.5, 11.5, 98)$	$(-1.5, 3.5, 101.3)$	13.2	13.7	**9.0**
(0, 5, 130)	$(-3, 4, 142)$	$(-2.5, 13.5, 130)$	$(0.5, 4.5, 126.3)$	12.4	8.9	**3.8**
(0, 5, 150)	$(-4, 5, 131)$	$(-3.5, 10.5, 123)$	$(-3.5, 4.5, 84.3)$	**19.4**	27.8	65.8
(13, −5, 90)	$(11, -8, 81)$	$(11.5, -4.5, 93)$	$(10.5, -7.5, 86.3)$	9.7	**3.4**	5.1
(13, −5, 110)	$(11, -7, 93)$	$(12.5, -0.5, 98)$	$(11.5, -7.5, 109.3)$	17.2	12.8	**3.0**
(13, −5, 130)	$(10, -8, 118)$	$(10.5, -4.5, 131)$	$(0.5, -7.5, 123.3)$	12.7	**2.7**	14.4
(15, −5, 150)	$(10, -8, 118)$	$(12.5, -4.5, 162)$	$(9.5, -8.5, 135.3)$	32.3	**12.0**	15.5
(−21, −12, 90)	$(-21, -12, 90)$	$(-22.5, -6.5, 90)$	$(-21.5, -11.5, 94.3)$	**0.0**	5.7	4.4
(−21, −12, 110)	$(-21, -11, 94)$	$(-21.5, -4.5, 92)$	$(-21.5, 9.5, 105.3)$	**16.0**	19.5	22.0
(−21, −12, 130)	$(-18, -17, 169)$	$(-24.5, -5.5, 126)$	$(-21.5, -12.5, 112.3)$	39.4	**8.4**	17.7
(−21, −12, 150)	$(-21, -14, 165)$	$(-21.5, -3.5, 119)$	$(-22.5, -11.5, 143.3)$	15.1	32.1	**6.9**

4   Discussion and Conclusion

We presented a method how to learn spatial statistics from modeled photon multi-pinhole collimator interaction using CNNs, and tested its performance on synthetic and real 2D activity images. Our approach offers advantages over compressed sensing and fingerprinting algorithms thanks to improved processing speed during measurement, less constraints on host memory and thus the ability to be used on embedded computing boards. As the real experimental data is in general more varied than the modeled data, the accuracy of the method is currently limited but shows its potential, which deserves further research.

Intraoperative orientation provided by audio-based gamma detectors limits the effectiveness of SNB. However, a more targeted SNB enables a more reliable post-operative histopathologic staging, and therefore a more effective analysis of potential tumor spreading. The described deep neural network approach could lead to an improved SNB procedure in general. Next steps involve the development of more complex phantoms, equipped with simultaneously radiating low-dose (kBq) ^{99m}Tc sources, to evaluate and validate the sensitivity and reconstruction quality of the approach under such conditions, which correspond more to an expected in vivo environment.

References

1. Borbón-Arce, M., et al.: An innovative multimodality approach for sentinel node mapping and biopsy in head and neck malignancies. Revista Espanola de Medicina Nuclear e Imagen Molecular **33**(5), 274–279 (2014)
2. Wendler, T., et al.: First demonstration of 3-D lymphatic mapping in breast cancer using freehand SPECT. Eur. J. Nucl. Med. Mol. Imaging **37**(8), 1452–1461 (2010)
3. Seppi, C., et al.: Compressed sensing on multi-pinhole collimator SPECT camera for sentinel lymph node biopsy. In: Descoteaux, M., Maier-Hein, L., Franz, A., Jannin, P., Collins, D.L., Duchesne, S. (eds.) MICCAI 2017. LNCS, vol. 10434, pp. 415–423. Springer, Cham (2017). https://doi.org/10.1007/978-3-319-66185-8_47
4. Nahum, U., Seppi, C., von Niederhäusern, P.A., Pezold, S., Haerle, S.K., Cattin, P.C.: Sentinel lymph node fingerprinting. Phys. Med. Biol. **64**(11), 16 (2019). Article: 115028
5. Mansour, H.: Beyond l1-norm minimization for sparse signal recovery. In: IEEE Statistical Signal Processing Workshop, SSP 2012, pp. 337–340 (2012)
6. Henrich, B., et al.: PILATUS: a single photon counting pixel detector for X-ray applications. Nucl. Instrum. Methods Phys. Res., Sect. A **607**(1), 247–249 (2009)

Neural Denoising of Ultra-low Dose Mammography

Michael Green[1]([✉]), Miri Sklair-Levy[2], Nahum Kiryati[3], Eli Konen[2], and Arnaldo Mayer[2]

[1] School of Electrical Engineering, Tel-Aviv University, Tel Aviv-Yafo, Israel
greenl@mail.tau.ac.il
[2] Diagnostic Imaging, Sheba Medical Center, Affiliated to the Sackler School of Medicine, Tel-Aviv University, Tel Aviv-Yafo, Israel
{miri.sklairlevy, eli.konen, arnaldo.mayer}@sheba.health.gov.il
[3] The Manuel and Raquel Klachky Chair of Image Processing, School of Electrical Engineering, Tel-Aviv University, Tel Aviv-Yafo, Israel
nk@eng.tau.ac.il

Abstract. X-ray mammography is commonly used for breast cancer screening. Radiation exposure during mammography restricts the screening frequency and minimal age. Reduction of radiation dose decreases image quality. Image denoising has been recently considered as a way to facilitate dose reduction in mammography without impacting its diagnostic value. We propose a convolutional locally-consistent non-local means (CLC-NLM) algorithm for ultra-low dose mammography denoising. The proposed method achieves powerful denoising while preserving fine details in high resolution mammography. Validation is performed using a dataset of 16 digital mammography cases (4-views each). Since obtaining true low-dose and high-dose mammogram pairs raises regulatory concerns, we applied the X-ray specific and validated method of Veldkamp *et al.* to simulate 90% dose reduction. The proposed algorithm is shown to compete favorably, both quantitatively and qualitatively, against state-of-the-art neural denoising algorithms. In particular, tiny micro-calcifications are better preserved using the proposed algorithm.

1 Introduction

Mammography has been used for breast cancer detection since the 1970s. While MRI and ultrasound scanning contribute to breast imaging, mammography is still the only imaging modality recommended for breast cancer screening in the US [1]. Using modern equipment, the typical radiation dose for two-view mammography is about 0.4 mSv per breast, comparable to 7 weeks of background radiation [2]. Radiation exposure in mammography is a clinical consideration, limiting screening frequency. Of special concern are younger women, those with large breasts requiring extra views [3] and BRCA mutation carriers. Reducing radiation exposure in mammography is therefore a real necessity.

Dose reduction in mammography leads to reduced image quality; significant dose reduction leads to loss of diagnostic value. Image denoising has been considered as a way

© Springer Nature Switzerland AG 2019
F. Knoll et al. (Eds.): MLMIR 2019, LNCS 11905, pp. 215–225, 2019.
https://doi.org/10.1007/978-3-030-33843-5_20

to facilitate dose reduction in mammography without impacting its diagnostic value. Mammogram denoising using classical image processing techniques is reviewed in [4].

While mammography analysis by neural networks is an active research field, neural networks have been rarely considered for mammogram denoising. Interestingly, it has just recently been shown that classical denoising using a Wiener filter improves automatic neural-based detection of calcifications [5]. In [6], a convolutional neural network (CNN) was proposed for the denoising of low-dose digital breast tomosynthesis (DBT) images. The network was trained and tested using two breast cadaver phantoms.

In current state-of-the-art of image denoising, nonlocal filters (e.g. [7, 8]), CNN-based filters (e.g. [9, 10]) as well as methods based on generative adversarial networks (GAN, e.g. [11, 12]) are among the most powerful techniques [13].

Enforcing local consistency improves the performance of nonlocal denoising [7]. Neural and nonlocal image denoising approaches were combined in [13–16]. In [14], the image is first denoised by the classical BM3D algorithm [8], then noisy patches are stacked together, based on similarity of their denoised versions, to form three dimensional blocks. The blocks are denoised by a network similar to [9] and the output is obtained by aggregation of the denoised patches. In [13], an iterative method that combines CNN-based and non-local-based filters is proposed. The algorithm is modular and can combine any pre-trained CNN with any existing non-local filter. Reference [15] presents a gradient-descent based blind non-local denoising algorithm using a neural network incorporating patch grouping. A neural nearest-neighbor block is proposed in [16]. It can be combined with existing CNN-based methods (e.g., [9]), outperforming other neural non-local denoising algorithms [13–15].

State of the art medical image denoising methods use GANs. In [11] an adversarial loss was combined with SSIM [17] and L_1 loss to create a structure-sensitive denoising algorithm for low-dose CT. A universal image-to-image algorithm was proposed in [12], showing state-of-the-art results in three medical image processing problems, including PET-CT denoising.

In this paper we propose a convolutional locally-consistent nonlocal means (CLC-NLM) denoising algorithm, with application in low-dose mammography. Our contributions are both at the algorithmic and diagnostic dimensions. In the proposed algorithm, we incorporate a novel fully trainable nonlocal denoising block built on top of the locally-connected convolution layers. Furthermore, by avoiding weight sharing, the locally connected layers learn to perform space dependent denoising, enforcing local spatial consistency. Based on these algorithmic developments, we demonstrate the possibility of performing low-dose mammography with only 10% of the radiation while preserving its diagnostic value.

2   Methods

The LC-NLM algorithm [7] is a non-neural patch-based image denoising method that has been successfully applied in CT denoising. The CLC-NLM algorithm proposed in this paper adopts the rationale of the LC-NLM algorithm, casting its principles as one aspect of a deep neural computational framework. We briefly review the LC-NLM algorithm before detailing the proposed CLC-NLM algorithm.

2.1 The LC-NLM Algorithm

The LC-NLM algorithm uses fast approximate nearest neighbors (ANN) to find the most similar high-SNR patch, in a purposely built database, for each noisy patch in the input image [7]. The denoised pixel value at location i, denoted by p_{d_i}, can be formulated as

$$p_{d_i} = S_{LC}\left(P_i^*, \hat{P}_i\right)^T V\left(\hat{P}_i\right) \tag{1}$$

Explicitly, p_{d_i} is the dot product between vector $S_{LC}\left(P_i^*, \hat{P}_i\right) \in R^K$ and vector $V\left(\hat{P}_i\right) \in R^K$, where K is the number of patches containing any pixel i, that is $K = N^2$ for $N \times N$ patches. The notation is explained with reference to Fig. 1, where $N = 3$.

Let i (red) be any pixel in the noisy image. Pixel i is viewed as an anchor pixel in an $N \times N$ patch P_i^* (green) in the noisy image. The anchor pixel i is arbitrarily set at the upper-left corner of the patch P_i. Let P_i denote the set of patches (blue) in the noisy image containing pixel i. Note that $P_i^* \in P_i$, i.e., P_i^* is a specific member of P_i. For each $P_i \in P_i$, there is a nearest neighbor $\hat{P}_i$ in a set of high-SNR patches. The set $\hat{P}_i$ consists of high SNR patches, each being the nearest neighbor to a low SNR patch in P_i. Vector $V\left(\hat{P}_i\right) \in R^K$ contains the value of each patch $\hat{P}_i \in \hat{P}_i$ at the location of noisy pixel i. $S_{LC}\left(P_i^*, \hat{P}_i\right) \in R^K$ is the vector of similarities between noisy patch P_i^* and each high-SNR patch $\hat{P}_i \in \hat{P}_i$ computed at their intersection (grey).

In essence, Eq. 1 sets the denoised pixel value p_{d_i} to be the weighted average of pixel values at the same location in high-SNR patches corresponding to the low-SNR patches containing pixel i.

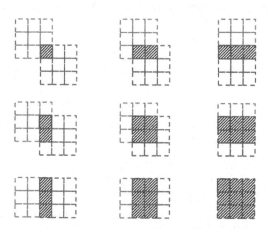

Fig. 1. LC-NLM geometry illustration for N = 3. Each noisy input patch P_i^* (green) has N^2 overlapping noisy patches P_i (blue) that contain anchor pixel i (red). Each of the overlapping patches P_i (blue) has a different overlap area (grey) with the input patch P_i^* (green). (Color figure online)

2.2   The Convolutional LC-NLM (CLC-NLM) Algorithm

The recently introduced Non-Local Neural networks (NLN) [18] use concepts of self-attention introduced in the context of natural language processing (NLP) in [19]. The output of the non-local block proposed in [18] can be written (omitting the residual connection) as

$$O = Z(S(x)G(x)) \tag{2}$$

where $x \in R^{mn}$ is the column-stacked version of the input $X \in R^{m \times n}$, $S(x) \in R^{mn \times mn}$ is a self-similarity matrix according to some similarity measure, meaning the value at location i,j equals $S(x)_{i,j} = \text{similarity}(x_i, x_j)$. $G : R^{mn} \to R^{mn}$ and $Z : R^{mn} \to R^{mn}$ are learned functions (e.g., convolutional layers).

Unlike in the original NLN, where similarity is computed between pixels, we compute similarity between patches. Specifically, similarity is computed between input patches extracted from noisy image X and their denoised versions, generated by a locally-connected CNN architecture [20]. The output of the non-local block becomes

$$O = Z(S(X^{ps}, G(X)^{ps})G(X)) \tag{3}$$

where $X^{ps} \in R^{mn \times N^2}$ is a patch-stacked version of X, where row i of X^{ps} is the row-stacked patch of size $N \times N$ extracted from X such that pixel i is in its upper left corner. $G(X)^{ps} \in R^{mn \times N^2}$ is the patch-stacked denoised version of X. $S(\cdot, \cdot) \in R^{mn \times mn}$ is the similarity matrix between each row of X^{ps} and each row in $G(X)^{ps}$, corresponding to any patch pair in X and $G(X)$. By converting X and $G(X)$ to their patch-stacked versions, the similarity matrix can be computed using simple matrix operators and seamlessly converted into a neural network layer. For the normalized cross-correlation similarity, $S_{NCC}(\cdot, \cdot)$ is given by

$$S_{NCC}(X^{ps}, G(X)^{ps}) = \frac{1}{N}\bar{X}\bar{G}(X)^T \triangleq \frac{1}{N}\left(\frac{X^{ps} - M_X}{STD_X}\right)\left(\frac{G(X)^{ps} - M_{G(X)}}{STD_{G(X)}}\right)^T \tag{4}$$

$M_X, STD_X, M_{G(X)}, STD_{G(X)}$ are the mean and standard deviation for each patch of X and $G(X)$, respectively. Exploiting the representation power of neural networks, we can improve the similarity measure using learned functions, θ, ϕ, leading to

$$S_{NCC}^l(X^{ps}, G(X)^{ps}) = \frac{1}{N}\theta(\bar{X})\phi(\bar{G}(X))^T \tag{5}$$

The learned functions θ, ϕ can be represented as 1×1 convolutions [18]. Likewise, we can generalize the SSIM [17] similarity measure to

$$S_{SSIM}^l(X^{ps}, G(X)^{ps}) = \frac{\theta_1\left(2M_X M_{G(X)}^T + C_1\right) \otimes \theta_2\left(2K_{XG(X)} + C_2\right)}{\theta_4(M_X^2 + M_{G(X)}^{2^T} + C_1) \otimes \theta_3(STD_X^2 + STD_{G(X)}^{2^T} + C_2)} \tag{6}$$

where C_1, C_2 are constants, $\otimes$ denotes element-wise multiplication, θ_k are learned functions (e.g., 1×1 convolutions), and $K_{XG(X)}$ is the correlation coefficient between X and $G(X)$, similar to S_{NCC}. The mean, standard deviation and the correlation coefficient are all computed using Gaussian weights [17]. In order to normalize the similarity matrix, i.e. $\sum_j S_{ij} = 1, 0 \le i < mn - 1$, the similarity matrix is forwarded through a softmax layer computed at each row.

Comparing the proposed non-local block to NLN [18], we note that while NLN is based on self-attention [19], we consider mutual-attention, i.e., attention between the input image and its denoised version. Furthermore, unlike [18], we use patches, perform denoising and propose a novel fully trainable structure similarity for the calculation of attention.

2.3   Enforcing Local-Consistency

In order to enforce locally consistent denoising, similarity computation is limited to patch pairs overlapping at the anchor pixel [7]. Non-local means denoising is associated with processing multiple hypotheses regarding the best denoised value at a given pixel location. Successful implementation of this concept call for a rich variety of hypotheses. We achieve this by preferring space-variance in the denoising network $G(\cdot)$, in contrast to the space-invariance property of common convolutional networks. In practice, we constructed $G(\cdot)$, with locally connected layers [20], which perform convolutions, but do not share the weights. Consequently, kernel parameters are different at each location.

In practice, we apply the locally-connected convolutional layers to a set of N^2 translated versions of X, denoted *trays* in the sequel, each corresponding to a different overlap configuration as shown for 3×3 patches in Fig. 2. Let $X_T \in R^{m \times n \times N^2}$ denote the trays, such that $X_T[:,:,k]$, with $0 \le k < N^2$, (blue) is the translation of X by $T_x = mod(k, N), T_y = \lfloor \frac{k}{N} \rfloor$, where $\lfloor \cdot \rfloor$ is the floor operator. The red pixel indicates the position of the anchor pixel for an arbitrary patch in the noisy input image (green). $G(X_T)^{ps} \in R^{mn \times N^2 \times N^2}$ can be conveniently reshaped to $R^{mnN^2 \times N^2}$, so that the denoising of tray $X_T[:,:,k]$ corresponds to rows, $mnk \le i < mn(k+1)$, in the reshaped $G(X_T)^{ps}$.

The proposed data structure enables direct computation of the similarity matrix, and the resulting CLC-NLM algorithm output, using Eqs. 6 and 3, respectively, with $X := X_T$. Furthermore, using the set of denoised trays, locally consistent denoising, that was enforced one pixel at a time in [7], has now become an efficient vectorized operation.

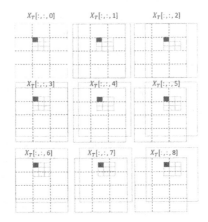

Fig. 2. The set X_T of N^2 translated versions (trays, solid blue) of noisy image X (solid green). Each tray in X_T corresponds to a different overlap configuration, shown here for 3×3 patches. The red pixel indicates the position of the anchor pixel for an arbitrary patch (green) in the noisy image X. For clarity, the subdivision of the trays is shown only at the patch level (dotted blue) but not to the pixel level as in the noisy (green) patch associated with the anchor pixel. (Color figure online)

3   Experiments

3.1   Dataset

A dataset of 16 retrospective digital mammography cases was used in the experiments, each containing the 4 standard mammographic views (L-CC, L-MLO, R-CC, R-MLO). The mammograms were all acquired on a Senographe Pristina (GE Healthcare, France) with about 7M pixels per view.

Ideally, experimental evaluation of ultra-low dose (ULD) mammogram-processing methods should include a dataset of true ultra-low dose (ULD) and corresponding normal dose mammogram pairs. However, obtaining such dataset requires doubling the number of X-ray images taken for each patient. While the additional radiation hazard due to the ULD acquisition is probably small, such experimental process cannot be carried out without a substantial risk analysis and ethical review process. Encountering a similar dilemma, Veldkamp *et al.* [21] developed and validated a specialized technique for simulating the effect of dose reduction on image quality in digital chest radiography. In this paper, we followed the principles of [21], but had to adapt their method for use in mammography.

Following [21], realistic ultra-low dose (ULD) mammograms, corresponding to a dose reduction of about 90%, were simulated by replacing the value of each pixel by a linear combination of its normal-dose value and random noise. The combination weights depend on the normal-dose value itself and the required dose reduction, leading to a highly nonlinear process. The mapping of pixel value to pixel-wise noise level was carried out in [21] by a lookup-table created using a dedicated chest phantom. For dose-reduction simulation in mammography, we used a synthetic breast phantom scanned with different radiation doses (Fig. 3). Specifically, the breast phantom was scanned at multiple radiation levels, and a small uniform ROI in the phantom was used to associate pixel value (mean of ROI values) with the corresponding noise level (via the standard deviation of ROI values).

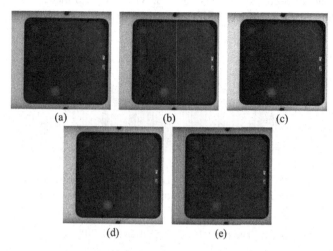

(a)   (b)   (c)

(d)   (e)

Fig. 3. A dedicated breast phantom scanned at different radiation doses (a–e)

3.2 Setup

In the neural implementation, convolution kernel size was set to 3×3. In the locally-connected layers the number of filters was set to 64 in all but the last the layer. Training was carried out using the common perceptual loss [22].

One case (4 views) was reserved for training while the remaining 15 were used for testing. No significant difference was observed when choosing a different scan for training. This may result from the very large number of patches extracted from each scan. The network was trained with 2×10^6 patch pairs (sampled at random from the 4 views) of size 35×35 with $N = 5$.

As shown in [23], common quality assessment indices like PSNR and SSIM do not always correspond to real visual quality. To overcome this limitation in a medical-imaging application, Perceptual Loss was proposed for the training of denoising neural networks [22]. Perceptual Loss enforces similarity between the network output and the ground-truth in a high dimensional feature space, rather than in image space. In [24] this idea was used to create a distance measure, called LPIPS, that is based on features extracted from networks pre-trained for image classification. Since LPIPS is a distance measure, where "smaller is better", to facilitate comparison with other quality indices, where "bigger is better", we use 1-LPIPS.

3.3   Comparison with State-of-the-Art

We compare the proposed algorithm to six state-of-the-art denoising algorithms: three CNN-based algorithms (DnCNN [9], WIN [10] and Perceptive [22]), a non-local neural algorithm (N3Net [16]), and two GAN-based algorithms (SMGAN [11] and MEDGAN [12]). Note that we used the 2D version of SMGAN, due to the 2D nature of mammograms. The PSNR, SSIM and LPIPS scores are shown in Fig. 4 for each case, and

Fig. 4. The PSNR, SSIM and LPIPS scores for each mammography case, denoised by DnCNN, WIN, Perceptive, N3Net, MEDGAN, SMGAN and the proposed algorithm.

Table 1. The mean PSNR, SSIM and 1-LPIPS scores.

	Our	DnCnn	WIN	Perceptive	N3Net	MEDGAN	SMGAN
PSNR	38.41	39.31	37.86	36.16	**41.64**	35.31	33.87
SSIM	**0.965**	0.946	0.940	0.949	0.952	0.947	0.933
1-LPIPS	**0.804**	0.623	0.618	0.771	0.745	0.702	0.610

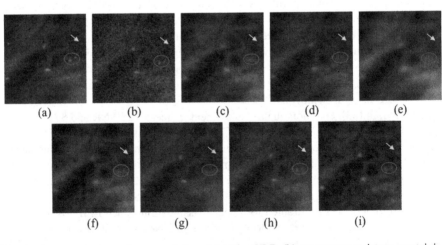

Fig. 5. A zoomed view of a normal-dose (a) and a ULD (b) mammogram image containing vacuum biopsy-proven micro-calcifications, with denoising results by DnCNN (c), WIN (d), SMGAN (e), MEDGAN (f), N3Net (g), Perceptive (h) and the proposed algorithm (i). Three tiny micro-calcifications (red circle, yellow arrow) are best visualized in our result (i). (Color figure online)

summarized in Table 1. The proposed algorithm achieved the highest mean SSIM and the highest mean LPIPS scores. In Fig. 5, zoomed views of a normal-dose (a) and ULD (b) mammogram containing vacuum biopsy-proven micro-calcifications are shown alongside the denoising results for DnCNN [9] (c), WIN [10] (d), SMGAN [11] (e), MEDGAN [12] (f), N3Net [16] (g), Perceptive [22] (h) and the proposed algorithm (i). Three tiny micro-calcifications (red circle, yellow arrow) are best visualized in our result (i).

3.4   Ablation Study

To further appreciate the contribution of the non-local block to the quality of the denoising results, we show in Fig. 6 (a) denoising of a sample tray by the locally connected network (i.e., $G(X_T[:, :, k])$), alongside the output of the consecutive non-local block using: (b) normalized cross-correlation similarity (S^l_{NCC}) and (c) SSIM similarity (S^l_{SSIM}). We note a considerable improvement in image quality between (a) and (b), provided by the non-local block. Also, the influence of the similarity function, used in the non-local block, is well visible: S^l_{SSIM} (c) provides better denoising and preservation of the tiny micro-calcifications (red circle, yellow arrow) than S^l_{NCC} (b).

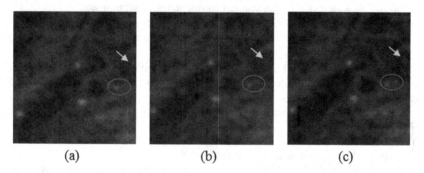

(a)   (b)   (c)

Fig. 6. (a) Denoising of a sample tray by the locally connected network (i.e., $G(X_T[:, :, k])$), alongside the output of the consecutive non-local block using: (b) normalized cross-correlation similarity (S^l_{NCC}) and (c) SSIM similarity (S^l_{SSIM}).

4   Conclusions

In this paper, a convolutional locally-consistent nonlocal means (CLC-NLM) algorithm was presented and applied to the challenging task of ultra-low dose (ULD) digital mammogram denoising. The fully trainable algorithm preserves important fine details such as micro-calcifications. The method was successfully validated against state-of-the-art neural denoising algorithms, using a realistic simulation of ULD mammograms, corresponding to 90% reduction in radiation dose. In an ongoing research project, the proposed method will be applied to larger datasets, including clinical ULD data.

References

1. U.S. Preventive-Services: Breast cancer: screening (2009). https://www.uspreventiveservicestaskforce.org/Page/Document/UpdateSummaryFinal/breast-cancer-screening
2. American Cancer Society: Mammogram basics (2017). https://www.cancer.org/cancer/breast-cancer/screening-tests-and-early-detection/mammograms/mammogram-basics.html
3. Miglioretti, D.L., et al.: Radiation-induced breast cancer incidence and mortality from digital mammography screening: a modeling study. Ann. Intern. Med. **164**(4), 205–214 (2016)
4. Vijikala, V., Dhas, D.A.S.: Identification of most preferential denoising method for mammogram images. In: Conference on Emerging Devices and Smart Systems (ICEDSS). IEEE (2016)
5. Marrocco, C., et al.: Mammogram denoising to improve the calcification detection performance of convolutional nets. In: 14th International Workshop on Breast Imaging (IWBI 2018). International Society for Optics and Photonics (2018)
6. Liu, J., et al.: Radiation dose reduction in Digital Breast Tomosynthesis (DBT) by means of deep-learning-based supervised image processing. In: Medical Imaging 2018: Image Processing. International Society for Optics and Photonics (2018)
7. Green, M., Marom, E.M., Kiryati, N., Konen, E., Mayer, A.: Efficient low-dose CT denoising by locally-consistent non-local means (LC-NLM). In: Ourselin, S., Joskowicz, L., Sabuncu, M.R., Unal, G., Wells, W. (eds.) MICCAI 2016. LNCS, vol. 9902, pp. 423–431. Springer, Cham (2016). https://doi.org/10.1007/978-3-319-46726-9_49
8. Dabov, K., Foi, A., Katkovnik, V., Egiazarian, K.: Image denoising by sparse 3-D transform-domain collaborative filtering. IEEE Trans. Image Process. **16**(8), 2080–2095 (2007)·
9. Zhang, K., Zuo, W., Chen, Y., Meng, D., Zhang, L.: Beyond a Gaussian denoiser: residual learning of deep CNN for image denoising. IEEE Trans. Image Process. **26**, 3142–3155 (2017)
10. Liu, P., Fang, R.: Wide inference network for image denoising. arXiv preprint arXiv:1707.05414 (2017)
11. You, C., et al.: Structure-sensitive multi-scale deep neural network for low-dose CT denoising. arXiv preprint arXiv:1805.00587 (2018)
12. Armanious, K., et al.: MedGAN: medical image translation using GANs. arXiv preprint arXiv:1806.06397 (2018)
13. Cruz, C., Foi, A., Katkovnik, V., Egiazarian, K.: Nonlocality-reinforced convolutional neural networks for image denoising. IEEE Signal Process. Lett. **25**(8), 1216–1220 (2018)
14. Ahn, B., Cho, N.I.: Block-matching convolutional neural network for image denoising. arXiv preprint arXiv:1704.00524 (2017)
15. Lefkimmiatis, S.: Universal denoising networks: a novel CNN architecture for image denoising. In: Proceedings of the IEEE Conference on Computer Vision and Pattern Recognition (2018)
16. Plötz, T., Roth, S.: Neural nearest neighbors networks. In: Advances in Neural Information Processing Systems (2018)
17. Wang, Z., Bovik, A.C., Sheikh, H.R., Simoncelli, E.P.: Image quality assessment: from error visibility to structural similarity. IEEE Trans. Image Process. **13**(4), 600–612 (2004)
18. Wang, X., Girshick, R., Gupta, A., He, K.: Non-local neural networks. In: Proceedings of the IEEE Conference on Computer Vision and Pattern Recognition (2018)
19. Vaswani, A., et al.: Attention is all you need. In: Advances in Neural Information Processing Systems (2017)

20. Taigman, Y., Yang, M., Ranzato, M.A., Wolf, L.: DeepFace: closing the gap to human-level performance in face verification. In: Proceedings of the IEEE conference on Computer Vision and Pattern Recognition (2014)
21. Veldkamp, W.J., Kroft, L.J., van Delft, J.P.A., Geleijns, J.: A technique for simulating the effect of dose reduction on image quality in digital chest radiography. J. Digit. Imaging **22** (2), 114–125 (2009)
22. Yang, Q., Yan, P., Kalra, M.K., Wang, G.: CT image denoising with perceptive deep neural networks. arXiv preprint arXiv:1702.07019 (2017)
23. Ledig, C., et al.: Photo-realistic single image super-resolution using a generative adversarial network. In: Proceedings of the IEEE conference on Computer Vision and Pattern Recognition (2017)
24. Zhang, R., Isola, P., Efros, A.A., Shechtman, E., Wang, O.: The unreasonable effectiveness of deep features as a perceptual metric. In: Proceedings of the IEEE Conference on Computer Vision and Pattern Recognition (2018)

Image Reconstruction in a Manifold of Image Patches: Application to Whole-Fetus Ultrasound Imaging

Alberto Gomez[1]([✉]), Veronika Zimmer[1], Nicolas Toussaint[1], Robert Wright[1], James R. Clough[1], Bishesh Khanal[1,2], Milou P. M. van Poppel[1], Emily Skelton[1], Jackie Matthews[1], and Julia A. Schnabel[1]

[1] Department of Biomedical Engineering, King's College London, London, UK
alberto.gomez@kcl.ac.uk
[2] NAAMII, Kathmandu, Nepal

Abstract. We propose an image reconstruction framework to combine a large number of overlapping image patches into a fused reconstruction of the object of interest, that is robust to inconsistencies between patches (e.g. motion artefacts) without explicitly modelling them. This is achieved through two mechanisms: first, manifold embedding, where patches are distributed on a manifold with similar patches (where similarity is defined only in the region where they overlap) closer to each other. As a result, inconsistent patches are set far apart in the manifold. Second, fusion, where a sample in the manifold is mapped back to image space, combining features from all patches in the region of the sample.

For the manifold embedding mechanism, a new method based on a Convolutional Variational Autoencoder (β-VAE) is proposed, and compared to classical manifold embedding techniques: linear (Multi Dimensional Scaling) and non-linear (Laplacian Eigenmaps). Experiments using synthetic data and on real fetal ultrasound images yield fused images of the whole fetus where, in average, β-VAE outperforms all the other methods in terms of preservation of patch information and overall image quality.

1 Introduction

Medical image reconstruction through fusion of partial captures consists of combining information from multiple images of the same object. Fusion is particularly useful when the images involved contain complementary information [7], for example when fusing Magnetic Resonance (MR) and Computed Tomography (CT) images of the brain [8] which shows more brain structures than any of the individual images, or when compounding multiple ultrasound (US) images of a

This work was supported by the Wellcome Trust IEH Award [102431]. The authors acknowledge financial support from the Department of Health via the National Institute for Health Research (NIHR) comprehensive Biomedical Research Centre award to Guy's & St Thomas' NHS Foundation Trust in partnership with King's College London and King's College Hospital NHS Foundation Trust.

F. Knoll et al. (Eds.): MLMIR 2019, LNCS 11905, pp. 226–235, 2019.
https://doi.org/10.1007/978-3-030-33843-5_21

fetus to provide whole body images [4,14]. The latter is the application targeted in this paper.

Fusion is normally a two step process: first, alignment of the images involved. Second, fusion of the aligned images. Alignment can be achieved by image registration [3]. Normally, rigid alignment is sufficient, if no motion is assumed between patches. In many cases, non-rigid motion can be expected, particularly in fetal imaging where the fetus moves frequently between acquisitions. Most research on image fusion has focused on discarding motion corrupted images or on correcting for motion using non-rigid registration [5,12]. However, registrations results are very sensitive to the registration method and the registration parameters. In the specific case of US imaging, motion correction using non-rigid registration can introduce visually abnormal patterns that degrade the quality of the reconstructed image. Moreover, the main cause of artefacts with state of the art methods is caused by motion and registration errors.

This paper introduces a novel and generic fusion framework for overlapping images (or patches) that have been aligned but may present residual registration errors and non-rigid motion artefacts. The aligned images are embedded into a manifold which separates motion corrupted patches, hence yielding a motion-free fused image without the need for non-rigid registration. The proposed method is evaluated on synthetic 2D images and on 3D fetal US.

2   Method

The key idea is illustrated in Fig. 1: we define a data set of $i = 1, \ldots, N$ image patches $I_i(\mathbf{x})$, spatially aligned (except for any non-rigid motion) and re-sampled into the same grid, so that the i-th patch only has information within a region defined by a binary mask $M_i(\mathbf{x})$. In this paper, patches are aligned rigidly using the method from [4]. Then, if patches i and j are similar in $M_i \cap M_j$, they are close neighbours in some manifold representation.

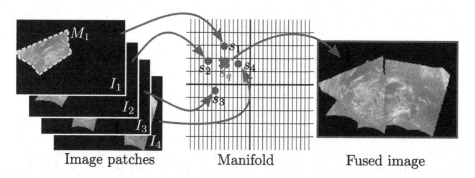

Image patches   Manifold   Fused image

Fig. 1. Overview of the method. Patches are embedded in a manifold, and the fused image can be retrieved by mapping a sample in the manifold back to image space.

Because the corresponding location s_i of I_i in the manifold represents the entire patch, if a query location s_q from the manifold can be projected back into the image space, then the resulting image has features from nearby patches, effectively fusing the data. In this paper we compare three methods to embed the patches into a manifold: linear embedding (multi dimensional scaling, MDS [11]), non-linear embedding (Laplacian Eigenmaps, LEM [1]) and variational autoencoders (β-VAE [6]).

2.1   Image Patch Fusion with Classical Manifold Embedding

Registered images capture aligned parts of the same fetus, and differences between them are due mainly to noise, artefacts, and possibly non-corrected motion. As a result, the main variation between images can be represented in a lower-dimensional manifold. Classical manifold embedding methods work by creating a neighbourhood graph between data points (patches), which is represented as a pair-wise resemblance matrix. Then, a linear or non-linear map $\mathcal{M} : \mathbb{R}^D \longrightarrow \mathbb{R}^d$ that minimizes the change of these distances and brings the data into a $d \ll D$-dimensional space (manifold) is computed. As a matter of fact, when a linear embedding is used, the result is equivalent to a weighted average of the most similar patches using MSE as similarity criteria.

We propose to compute the pair-wise resemblance $R_{i,j}$ only in the region where the pair of patches overlap, this is $R(I_i, I_j) = r(I_i \cap M_j, I_j \cap M_i)$, where r is, here, the mean square error (MSE). This enforces that consistent patches are clustered together in the manifold, and indeed if two different patches I_i and I_j are identical in the region where they overlap, then $\mathcal{M}(I_i) = \mathcal{M}(I_j) = s \in \mathbb{R}^d$. If we could compute the inverse mapping $F = \mathcal{M}^{-1}(s)$, then F would fuse the information of I_i and I_j. However, most manifold embedding techniques are not invertible, so $\mathcal{M}^{-1}$ cannot be computed. We can estimate the fused image $\mathcal{F}(s_q)$ corresponding to a query sample s_q in the manifold by interpolating nearby samples, e.g. using Shepard's interpolation [13]:

$$\mathcal{F}(s_q) = \frac{\sum_{i \in \Omega_q} I_i M_i w_i(s_q)}{\sum_{i \in \Omega_q} M_i w_i(s_q)} \qquad (1)$$

where Ω_q is a neighbourhood on the manifold around s_q and w_i is the distance-based Shepard's weight, also computed on the manifold as $w_i(s_q) = 1/\|s_q - s_i\|^2$.

2.2   Image Patch Fusion with a Variational Autoencoder

Autoencoders encode input data into a lower dimensional (latent) space, with the advantage that they provide a decoder sub-net to map the latent space back into the original space, effectively implementing the sought $F = \mathcal{M}^{-1}(s)$ mapping. β-VAEs [6,10] additionally constrain the latent space to be normally distributed, which produces consistent images from the entire manifold as the latent space is continuous by construction. As a result, β-VAEs are ideally placed to build the manifold from patch images and to retrieve a fused image from the query manifold sample s_q. Normally, VAEs (and more broadly, neural-networks) are used to learn models of a population, and then make some predictions from unseen data. Crucially, in this work we propose to learn a model of specific subject (fetus), of which we have a large number of partial observations (overlapping patches), which yields not a population model, but a reconstruction of the subject itself. Then, instead of querying the model to obtain predictions of unseen data, the latent space (manifold) can be queried to reconstruct different poses of the subject.

We assume that each input patch $\{I_i\}_{i=1}^{N}$, will be similar (in the least squares sense, and within M_i) to the fused image $F_i = F(s_i)$ except for a normally distributed random noise, i.e. $I_i - F_i \cap M_i = \epsilon_i \sim \mathcal{N}(0, \sigma^2)$. As a result, $P(F|s) = P(\epsilon|s) = P(I|s)$. In consequence, the log likelihood estimator of $P(F|s)$ yields the MSE:

$$\log P(F|s) = \sum_{i=1}^{N} \log P(I_i - F_i \cap M_i|s) = C + \sum_{i=1}^{N} \frac{||I_i - F_i(s) \cap M_i||^2}{2\sigma^2} \quad (2)$$

Noting the encoder function $s = f_\phi(I)$, and adding the Kullback-Leiber (KL) divergence (weighted by β [6], which allows a trade-off between data fidelity and normal distribution of the latent space) the loss becomes:

$$\mathcal{L}(\theta, \phi, \beta; \{I\}, \{s\}) = \sum_{i=1}^{N} ||I_i - F_\theta(f_\phi(I_i)) \cap M_i||^2 + \beta KL(s_i) \quad (3)$$

The fused image can be reconstructed by sampling the latent space at $\{s_q\}$.

3 Materials and Experiments

3.1 Materials

We carry out experiments on a synthetic and real data-sets. The synthetic data-set consisted on 8 images of 128×128 pixels illustrating a fetus where the leg was at different locations as if captured during a kick. These images were divided into 280 overlapping patches of 40×40 pixels, with a 70% overlap both vertically and horizontally, to which Gaussian noise $\sim \mathcal{N}(\mu = 0, \sigma = 5)$ was added.

Experiments using 3D and 2D images from healthy fetal subjects were carried out. 3D ultrasound image sequences were acquired from two fetuses (GA 32w, 24w). Patient 1 was acquired over a head to toe sweep in which 120 volumes were acquired. Patient 2 was acquired over 5 consecutive sweeps head to toe and back, totalling 470 volumes. In both cases, data was acquired using a Philips EPIQ system with a X6-1 transducer at 4 volumes/s.

Ultrasound data was registered using the method from [4], using a grid of 1500 points distributed evenly, and each registered image was transformed and re-sampled into the fusion space at $1\,\text{mm}^3$, totalling $181 \times 95 \times 226$ and $172 \times 175 \times 185$ voxels per volume for each patient. Registered input volumes were sliced through a longitudinal plane to produce 2D patches (275×184 and 352×285 pixels, respectively), which were used for the 2D experiments on real data. The β-VAE architecture was inspired by the 3D branch in [2] and represented in Fig. 2. The kernel size in all conv layers is 3×3 ($\times 3$ in 3D). Training was carried out using the Adam optimizer [9] with a learning rate of 10^{-4} and over 300 epochs.

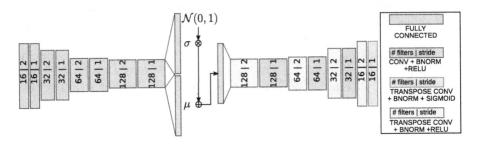

Fig. 2. Architecture of the VAE. Convolution kernels are of size 3×3 ($\times 3$ in 3D).

3.2 Experiments

Whole-body fusion of the 2D fetal ultrasound images was used for quantitative and qualitative validation. Quantitative evaluation was aimed at establishing the ability of the method to get rid of fusion artefacts (namely non-rigid motion and registration errors) while capturing the whole fetal anatomy, through the quality metric $Q(i,j) = \sqrt{\frac{1}{|M_i|} \sum_{x \in M_i} ||I_i(x) - F_j(x)||^2}$, so that $Q_{IN}(i) = Q(i,i)$, measures the RMS difference between an input patch and the same region in the fused image F_i reconstructed from the corresponding sample s_i in the manifold, therefore it measures to what extent the information in the patch was preserved. In order to measure the quality of the fusion outside the input patch I_i, we use $Q_{OUT}(i) = \frac{1}{|\Omega|} \sum_{j \in \Omega} Q(i,j)$, where the set Ω is built incorporating the patch $I_j, j \neq i$, in increasing order $Q(i,j)$, that does not intersect with patch I_i or with any of the patches already in Ω.

Further qualitative evaluation was conducted on the experiments by measuring the subjective appearance of the fused images. Three raters were presented with 500 pairs of fused images, randomly selected from a uniform sampling grid in the manifold. The raters were asked to select which image was best, or if they were of equal quality.

Both qualitative and quantitative evaluations were carried out on the synthetic and real datasets using the three methods: linear and non-linear manifold embedding and β-VAE fusion.

4   Results

The quantitative results are provided in Table 1. The pixel-wise average fusion (Avg.) was used as baseline, and compared to the two manifold embedding methods (MDS and LEM) and to the β-VAE fusion. β values in the range $[1 \cdot 10^{-4}, 1 \cdot 10^{-2}]$ were used, and a subset of this range, where results were best ($\beta \in [3 \cdot 10^{-4}, 2 \cdot 10^{-3}]$) is reported. The rows where values of β yield significantly worse results have been greyed out, so the remaining values give an idea of the range of β where results are stable. Lower values of β introduced an increasing amount of noise in the reconstruction, and higher values introduced blur, as the latent space collapsed into a single point for $\beta > 0.01$. The results show that the β-VAE approach outperforms all the other in preserving the features from individual patches. $\beta = 7 \cdot 10^{-4}$ was used for the qualitative experiments.

Table 1. Quantitative results on quality of the fused 2D image, measuring the ability of the method to preserve features from the input patches and provide a whole-fetus fusion, reported for the synthetic dataset (Synth.) and patients 1 and 2 (P1 and P2). Best values are highlighted in bold, worst are greyed out. All manifold fusion implementations outperform the naive average fusion, with the non-linear embedding (LEM) being the worst (interpolation brings the fused image outside the manifold). Overall, the β-VAE performs best, stable over a range of beta values. For patient 1, where the patch alignment is particularly good and the non-rigid motion limited over a small region covering the forearm, inter-method differences are less obvious.

| | | Q_{IN} | | | Q_{OUT} | | |
		Synth.	P1	P2	Synth	P1	P2
Avg		143.4 ± 33.8	153.8 ± 3.7	111.4 ± 10.6	112.9 ± 15.2	147.3 ± 3.8	96.7 ± 6.5
MDS		16.6±16.4	45.2±6.0	48.6±7.8	6.8±5.7	25.7±9.6	32.1±9.9
LEM		17.6±17.9	47.1±7.1	50.1±8.3	6.3±5.5	25.3±10.5	32.1±9.7
	β						
β-VAE	3E-4	14.1±16.4	**24.4±2.4**	17.4±2.7	6.1±3.4	**23.7±4.8**	46.6±23.5
	4E-4	14.7±17.5	25.9±2.9	**17.3±2.5**	6.4±4.0	37.0±12.5	38.5±20.2
	5E-4	13.4±15.3	25.3±3.0	17.5±2.4	6.5±3.0	24.9±5.1	44.2±23.7
	6E-4	13.6±16.5	25.6±2.7	17.8±2.9	6.1±3.6	26.4±5.3	38.8±22.1
	7E-4	**12.4±14.1**	26.4±3.1	17.8±2.8	6.2±3.1	27.9±6.7	**38.1±18.0**
	8E-4	14.2±16.3	27.2±3.1	18.0±2.9	**6.0±3.0**	30.0±6.8	40.8±21.6
	9E-4	13.6±15.1	27.7±3.1	18.2±3.0	6.8±3.8	27.5±5.0	36.4±19.5
	1E-3	13.8±15.7	28.0±3.6	18.7±3.0	6.7±4.3	29.4±6.8	39.6±19.1
	2E-3	13.0±14.3	32.5±4.2	19.6±3.3	6.3±3.6	31.6±4.9	29.7±14.4

Qualitative results in Fig. 3, show the amount of images (in %) where the fusion using the β-VAE method was judged better than the other methods, for each data-set. Overall, the β-VAE method provided better fusions with less artefacts. In the case of P1, there is no motion artefacts except for the fetal arm (c.f. second row in Fig. 4), therefore the average reconstruction is of high quality already. This explains the difference with the other data-sets.

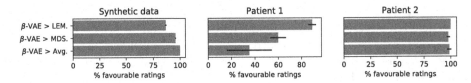

Fig. 3. β-VAE > Y indicates fraction of the times (in %) where the results with the β-VAE where considered better than with method Y by 3 raters. The bars show the average and standard deviation. The β-VAE outperforms all the other methods, with the exception of P1 where the average fusion is chosen best more often. As supported by Table 1, P1 combined good patch alignment and limited motion artefacts, so the smooth appearance of the average fusion was found to be visually best.

Examples of representative 2D fusions are shown in Fig. 4, where the fused images mapped back from the manifold sample corresponding to the patch on the left column is shown. The ability of the β-VAE to provide reconstructions without motion/blur artefacts is pointed at with white arrows. For example the second row shows, for patient 1, the β-VAE reconstructions generated from different manifold locations (corresponding to input patches marked by the red contour) that recover the entire fetus but with the arm on a different pose.

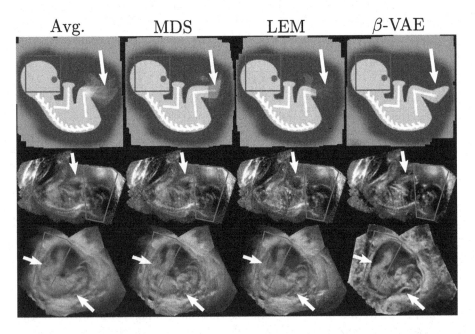

Fig. 4. 2D Whole-fetus fusions (for synthetic data, patient 1 and 2, in rows 1, 2 and 3 respectively), obtained from sampling the manifold at the location corresponding to one of the input patches. The region covered by the patch is outlined in red. White arrows indicate regions where fusion is challenging due to motion or mis-registration in input data. (Color figure online)

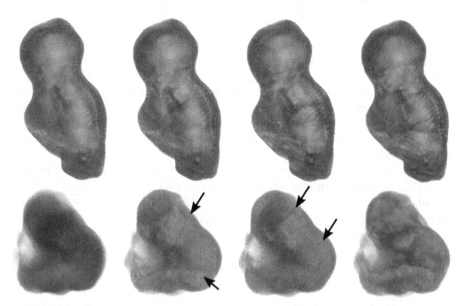

Fig. 5. Example of 3D fusion of patients 1 (top, all methods provide similar visual results) and 2 (bottom). Fusion methods, from left to right: average, MDS, LEM and β-VAE. The arrows point at some of the artefacts that were found systematically with classic manifold embedding techniques.

As a proof-of-concept, examples of the 3D version of the four methods for patients 1 and 2 are shown in Fig. 5 (as anticipated in the 2D experiments, volume renders of 3D reconstructions for patient 1 are indistinguishable). This result shows the potential of the proposed method to reconstruct high quality whole-body fetal images even from motion-corrupted input data, which would otherwise blur the result.

5 Discussion and Conclusions

We have presented a new paradigm to carry out fusion of a large amount of image patches, based on embedding the patches into a manifold through a map $\mathcal{M}$, and then sampling the manifold to reconstruct a fused image in the input image space. The proposed paradigm has been implemented using Multi-Dimensional Scaling, Laplacian Eigenmaps and a β variational autoencoder.

The inverse mapping $\mathcal{M}^{-1}$ was not available for classic manifold embedding techniques (e.g. MDS, LEM) and the fused image was obtained by Shepard's interpolation of input patches that are nearby in the manifold. Although non-linear embeddings potentially yield more accurate representation of the data, recovering the fused image through interpolation results in out-of-manifold images, which is why LEM produced worse results than the other methods. This is, to the best of our knowledge, the first time that fusion has been approached as learning a single-instance model, where all the data are of the same object, as opposed to the common practice of creating a model that captures the variability of a population. One advantage is that this eliminates bias from under-represented cases in a training set, e.g. rare morphological abnormalities.

A limitation of the proposed framework is that it does not distinguish between inter-patch differences due to motion, noise, etc. This lends itself towards disentangled representation of these sources of variation, particularly since it may be desirable to average over noise and artefacts while separating motion. This will be investigated in future work.

The proposed method shows promising results on image fusion of rigidly pre-aligned image patches, and particularly towards a challenging task as whole body fetal image fusion. The fused images reduce the artefacts caused by non-rigid motion and misalignment by pushing the problematic patches to relatively far regions in the manifold.

References

1. Belkin, M., Niyogi, P.: Laplacian eigenmaps for dimensionality reduction and data representation. Neural Comput. **15**(6), 1373–1396 (2003)
2. Cerrolaza, J.J., et al.: 3D fetal skull reconstruction from 2DUS via deep conditional generative networks. In: Frangi, A.F., Schnabel, J.A., Davatzikos, C., Alberola-López, C., Fichtinger, G. (eds.) MICCAI 2018. LNCS, vol. 11070, pp. 383–391. Springer, Cham (2018). https://doi.org/10.1007/978-3-030-00928-1_44
3. Che, C., Mathai, T.S., Galeotti, J.: Ultrasound registration: a review. Methods **115**, 128–143 (2017)
4. Gomez, A., Bhatia, K., Tharin, S., Housden, J., Toussaint, N., Schnabel, J.A.: Fast registration of 3D fetal ultrasound images using learned corresponding salient points. In: Cardoso, M.J., et al. (eds.) FIFI/OMIA -2017. LNCS, vol. 10554, pp. 33–41. Springer, Cham (2017). https://doi.org/10.1007/978-3-319-67561-9_4
5. Heinrich, M.P., Jenkinson, M., Papież, B.W., Brady, S.M., Schnabel, J.A.: Towards realtime multimodal fusion for image-guided interventions using self-similarities. In: Mori, K., Sakuma, I., Sato, Y., Barillot, C., Navab, N. (eds.) MICCAI 2013. LNCS, vol. 8149, pp. 187–194. Springer, Heidelberg (2013). https://doi.org/10.1007/978-3-642-40811-3_24
6. Higgins, I., et al.: β-VAE: learning basic visual concepts with a constrained variational framework. In: ICLR, pp. 1–22 (2017)
7. James, A.P., Dasarathy, B.V.: Medical image fusion: a survey of the state of the art. Inf. Fusion **19**(1), 4–19 (2014)
8. Kavitha, C.T., Chellamuthu, C.: Multimodal medical image fusion based on integer wavelet transform and neuro-fuzzy. In: IEEE ICSIP, pp. 296–300 (2010)
9. Kingma, D.P., Ba, J.: Adam: a method for stochastic optimization. In: ICLR (2015)
10. Kingma, D.P., Welling, M.: Auto-Encoding Variational Bayes. In: ICLR, December 2014
11. Mead, A.: Review of the development of multidimensional scaling methods. Statistician **41**(1), 27 (1992)
12. Rivaz, H., Chen, S.J.-S., Collins, D.L.: Automatic deformable MR-ultrasound registration for image-guided neurosurgery. IEEE TMI **34**(2), 366–380 (2015)

13. Shepard, D.: A two-dimensional interpolation function for irregularly-spaced data. In: Proceedings of ACM, pp. 517–524 (1968)
14. Wachinger, C., Wein, W., Navab, N.: Three-dimensional ultrasound mosaicing. In: Ayache, N., Ourselin, S., Maeder, A. (eds.) MICCAI 2007. LNCS, vol. 4792, pp. 327–335. Springer, Heidelberg (2007). https://doi.org/10.1007/978-3-540-75759-7_40

Image Super Resolution via Bilinear Pooling: Application to Confocal Endomicroscopy

Saeed Izadi$^{(\boxtimes)}$, Darren Sutton, and Ghassan Hamarneh

School of Computing Science, Simon Fraser University, Burnaby, Canada
{saeedi,darrens,hamarneh}@sfu.ca

Abstract. Recent developments in image acquisition literature have miniaturized the confocal laser endomicroscopes to improve usability and flexibility of the apparatus in actual clinical settings. However, miniaturized devices collect less light and have fewer optical components, resulting in pixelation artifacts and low resolution images. Owing to the strength of deep networks, many supervised methods known as super resolution have achieved considerable success in restoring low resolution images by generating the missing high frequency details. In this work, we propose a novel attention mechanism that, for the first time, combines 1st- and 2nd-order statistics for pooling operation, in the spatial and channel-wise dimensions. We compare the efficacy of our method to 10 other existing single image super resolution techniques that compensate for the reduction in image quality caused by the necessity of endomicroscope miniaturization. All evaluations are carried out on three publicly available datasets. Experimental results show that our method can produce superior results against state-of-the-art in terms of PSNR, and SSIM metrics. Additionally, our proposed method is lightweight and suitable for real-time inference.

1 Introduction

Colorectal cancer is known as the fourth most-common cancer and remains one of the leading causes of cancer related mortality in the world. In 2018, more than 1 million people were affected by colorectal cancer worldwide, resulting in an estimated 550,000 deaths [2]. Rapid histopathologic assessment is an important tool that may improve disease prognosis by detecting early-stage cancer and precancerous conditions. Although biopsy and *ex-vivo* tissue examination are widely accepted as the diagnostic gold standard, such procedures take time and may limit the ability of the endoscopist to rapidly gauge disease severity. Confocal laser endomicroscopy (CLE), on the other hand, has substantially improved real-time *in-vivo* visualization of the subsurface of living cells, vascular structures, and tissue patterns during endoscopic examination [10].

For *in-vivo* histological examination, the large size of the microscope complicates navigation of the interior of the body in a clinical setting. Therefore,

F. Knoll et al. (Eds.): MLMIR 2019, LNCS 11905, pp. 236–244, 2019.
https://doi.org/10.1007/978-3-030-33843-5_22

it is necessary to reduce the size of the microscope to completely and safely access the organ(s) of interest. However, miniaturization reduces the number of optical elements in the microscope probe, introducing pixelation artifacts in the acquired images. One strategy to remove image artifacts and enhance image quality is to directly post-process degraded images. An efficient process in the field of image processing, referred to as single image super-resolution (SR), aims to reconstruct an accurate high-resolution (HR) image given its low-resolution (LR) counterpart. Thus, SR is a promising software method to mitigate image degradation due to hardware miniaturization.

Among traditional SR algorithms, Huang et al. [8] proposed leveraging self-similarity modulo affine transformations to accommodate natural deformation of recurring statistical priors within and across scales of an image. Timofte et al. [19, 20] used a combination of neighbour embedding and sparse dictionary learning over an external database and proposed anchored neighborhood regression in the dictionary atom space. Recently, CNNs have advanced the SR research field by directly learning the mapping between LR and HR images [1,4,11–13]. Dong et al. [4] demonstrated that a fully convolutional network trained end-to-end can perform LR-to-HR nonlinear mapping. Kim et al. [11] suggested a trained network to predict additive details in the form of a residual image, which is summed with the interpolated image. Kim et al. [12] addressed model overfitting by reducing the number of parameters via recursive convolutional layers. Lai et al. [13] designed a network which progressively reconstructs the sub-band residuals of high-resolution images at multiple pyramid levels. Ahn et al. [1] improved speed and efficiency of SR models by designing a cascade mechanism over residual networks. Lastly, Cheng et al. [3] exploited recursive squeeze and excitation modules in a network to exploit relationships between channels. Izadi et al. [9] reported the first attempt to deploy CNNs on CLE images. They used a densely connected CNN to transform synthetic LR images into HR ones. Ravi et al. [15] employed a CNN to restore missing details into LR images. They collected a set of consecutive LR frames and generated synthetic HR images using a video registration technique. In a more recent study [16], Ravi et al. trained a CNN for unsupervised SR on CLE images using a cycle consistency regularization, designed to impose acquisition properties on the SR images.

In this paper, we present a lightweight convolutional neural network (CNN) that is appropriate for frame-wise SR by incorporating a novel attention mechanism. In contrast to SESR [3], which leverages attention modules from the Squeeze-and-Excitation network (SENet) [7] to re-weight channels, we introduce a novel weighting scheme to recalibrate learned features based on pairwise relationships. Our attention modules compromise both 1st-order pooling and 2nd-order pooling (a.k.a. bilinear pooling), improving the quality of learned features in the network by considering pairwise correlations along feature channels and spatial regions [5]. The compactness and computational speed of our network lends well to real-time implementation during *in-vivo* examination.

We demonstrate that stacking attention modules in the middle of a low-level feature extraction head and a feature integration tail quantitatively and qualitatively produces superior results against existing SR methods and generalizes well over unseen microscopic datasets.

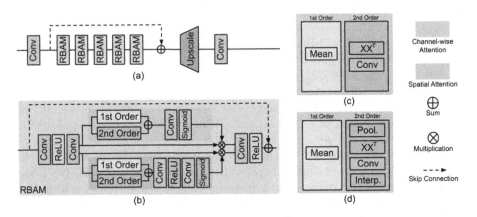

Fig. 1. (a) The overall architecture of our proposed network. (b) RBAM architecture. (c) channel-wise and (d) spatial attention architectures.

2   Method

Network Overview. Figure 1-a depicts the overall architecture of our proposed LR-to-SR network. Let $L_{LR} \in \mathbb{R}^{1 \times W \times H}$, $I_{SR} \in \mathbb{R}^{1 \times rH \times rW}$, and r denote the low resolution input and super-resolved output, and the downsample factor, respectively. We use a convolution layer, denoted by $\mathcal{F}(\cdot)$, with a 3×3 kernel and C output channels to extract initial features $H^0 \in \mathbb{R}^{C \times H \times W}$, i.e.

$$H^0 = \mathcal{F}_c(I_{LR}; \theta_0), \tag{1}$$

where θ refers to the learnable parameters. In our proposed network, the initial features H^0 are updated by sequential residual attention modules, denoted as $\mathcal{G}(\cdot)$ and a skip connection. The entire high-level feature extraction stage is denoted as $\mathcal{B}(\cdot)$:

$$H^{\mathcal{B}} = \mathcal{B}(H^0) = \mathcal{G}^b(\mathcal{G}^{b-1}(\dots(\mathcal{G}^1(H^0))\dots)) + H^0. \tag{2}$$

To upsample the feature maps, we use sub-pixel convolutions [17], denoted as $\mathcal{U}(\cdot)$, followed by a single channel 1×1 convolution for SR reconstruction:

$$I_{SR} = \mathcal{F}_1(\mathcal{U}(H^{\mathcal{B}}; \theta_{up}); \theta_{rec}). \tag{3}$$

Residual Bilinear Attention Module. In our proposed RBAM, we combine 1st- and 2nd-order pooling operations spatially and channel-wise to recalibrate learned features for efficient network training. Figure 1-b illustrates the structure of our proposed RBAM. Mathematically, we formulate RBAM as:

$$H^b = \mathcal{G}^b(H^{b-1}) = \mathcal{Q}^b(H^{b-1}) + H^{b-1}, \tag{4}$$

where $\mathcal{Q}(\cdot)$ denotes the attention modules before the skip connection. Given the input feature maps $H^b \in \mathbb{R}^{C \times H \times W}$, two convolutions with 3×3 kernel size interleaved with a ReLU activation function, denoted by $^+$, produce high-level feature maps $H^b_{conv} \in \mathbb{R}^{C \times H \times W}$ as input to the attention branches:

$$H^b_{conv} = \mathcal{F}_c(\mathcal{F}_c^+(H^{b-1}; \theta_1^b); \theta_2^b). \tag{5}$$

Channel-wise Attention (CA) Branch. CA leverages the inter-channel correspondence between feature responses (Fig. 1-c). 1st- and 2nd-order pooling mechanisms operate on H^b_{conv}, producing two vectors F^{1st}_{ca}, $F^{2nd}_{ca} \in \mathbb{R}^{C \times 1 \times 1}$. F^{1st}_{ca} is the 1st-order CA obtained by spatial average pooling to squeeze the feature map of each channel [7]. To obtain 2nd-order CA, pairwise channel correlations are computed in the form of a covariance matrix $\Sigma \in \mathbb{R}^{C \times C}$ by spatial flattening, dimension permutation, and matrix multiplication. Each row in Σ encodes the statistical dependency of a channel with respect to every other channel [5]. Given the covariance matrix Σ, we adopt a row-wise convolution with $1 \times C$ kernel size to produce the 2nd-order CA vector F^{2nd}_{ca}. Finally, two successive 1-D convolutions interleaved with a ReLU activation function, operate on a vector formed by the sum of $F^{1st}_{ca} + F^{2nd}_{ca}$. The output of the convolution operation is fed into a sigmoid function σ, followed by element-wise multiplication $\otimes$ to produce the bth updated features maps H^b_{ca}:

$$H^b_{ca} = H^b_{conv} \otimes \sigma(\mathcal{F}_c(\mathcal{F}_{\frac{c}{4}}^+(F^{1st}_{ca} + F^{2nd}_{ca}; \theta_3^b); \theta_4^b)). \tag{6}$$

Spatial Attention (SA) Branch. SA indicates shared correspondence between spatial regions across all feature maps (Fig. 1-d). Given H^b_{conv} as the input, the 1st-order spatial attention matrix, $F^{1st}_{sa} \in \mathbb{R}^{1 \times H \times W}$, is computed by the average pooling operation along channel dimension to aggregate information for each spatial location across all features. To compute 2nd-order spatial attention matrix, $F^{2nd}_{sa} \in \mathbb{R}^{1 \times H \times W}$, we first reduce the spatial size of feature maps to $H' \times W'$ (8×8 in our implementation) by applying average pooling. Then, appropriate reshaping, dimension permutation and matrix multiplication is adopted to obtain the covariance matrix $\Sigma \in \mathbb{R}^{H'W' \times H'W'}$. Similar to channel-wise attention, a row-wise convolution with $1 \times H'W'$ kernel size is applied on Σ. Eventually, dimension permutation and nearest neighbor interpolation produce F^{2nd}_{sa}. We add these two matrices together element-wise and apply a convolution with 1×1 kernel size that feeds a sigmoid function. Spatial attention is realised by element-wise multiplication over all feature maps, formulated as:

$$H^b_{sa} = H^b_{conv} \otimes \sigma(\mathcal{F}_c(F^{1st}_{sa} + F^{2nd}_{sa}; \theta_5^b)) \tag{7}$$

Attention Fusion. The updated features are concatenated ($+\!\!\!+$) and aggregated via a convolution with kernel 1×1 kernel, followed by a ReLU function. Lastly, H^b is added via skip connection:

$$H^b = \mathcal{F}_c^+(H^b_{ca} +\!\!\!+ H^b_{sa}; \theta_6^b) + H^{b-1}. \tag{8}$$

Table 1. Details of the datasets used in our evaluation.

dataset	provided by	#patients	#images	anatomical site	image size
CLE100	Leong et al. [14]	30	181	small intestine	1024 × 1024
CLE200	Grisan et al. [6]	32	262	esophagus	1024 × 1024
CLE1000	Ştefănescu et al. [18]	11	1025	colorectal mucosa	1024 × 1024

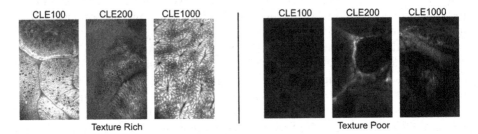

Texture Rich   Texture Poor

Fig. 2. (a) Examples of images from the partitioned test set. Images belonging to the 'texture rich' partition are used for evaluation.

3   Results and Discussion

Data. We evaluate existing state-of-the-art SR methods, as well as our proposed RBAM, on three publicly available CLE datasets (Table 1). We select images rich in texture by assessing the SR performance of bicubic interpolation on the unseen test set. As depicted in Fig. 2, images with PSNR scores below the mean PSNR score of the bicubic method evaluated on the test set are deemed 'texture rich', and are used for evaluation, whereas images associated with scores above the mean are deemed 'texture poor'. In other words, images which can be effectively restored using bicubic interpolation are rejected for evaluation, as they contain little information on which to assess the performance of state-of-the-art methods. Evaluation assesses the methods' ability to reconstruct 1024 × 1024 HR image from a synthesized LR counterpart obtained via bicubic downsampling with the appropriate factor (×2 or ×4).

Training Settings. We train all methods on a random partition (80%) of CLE100 as this dataset is the richest in texture. Methods are evaluated on the remaining 20% of CLE100, and all of CLE200 and CLE1000. For DL-based methods, we replicated the reported training settings, and used public code for traditional algorithms. For our model, we use $B = 5$ RBAMs and set the number of features to $C = 64$ to create a lightweight network. In each training batch, 16 LR patches of size 48 × 48 are randomly extracted as inputs, and augmented by random 90° rotations and horizontal/vertical flip. We use Adam optimizer and L1 loss to train our network for 300 epochs. Initial learning rate is set to 10^{-4} and is halved every 50 epochs.

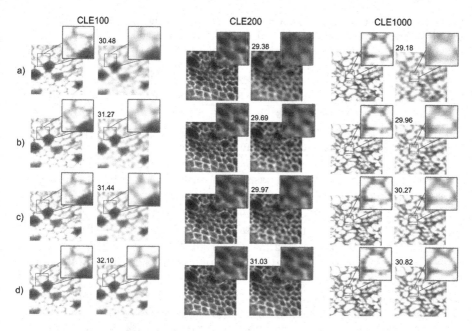

Fig. 3. Qualitative results and their PSNR scores at ×4 SR. Each row shows the side-by-side comparison of HR with (a) bicubic, (b) GR, (c) SESR, and (d) RBAM across three datasets. HR images are shown for each pair for ease of visual comparison.

Ablation Investigation. We discern the effectiveness of the individual components in our network modules by ablating attention blocks and evaluating performance after 50 epochs. Our investigation shows that, for CLE100 at ×2 SR, attention-based variants outperform the baseline, demonstrating the merits of incorporating spatial and channel-wise contextual information. We also observed that using both 1st and 2nd-order pooling operations simultaneously outperform using either 1st or 2nd-order channel-wise pooling individually. We similarly note that using both spatial and channel-wise attention outperforms either one alone.

Comparison to State-of-the-art. We compare the performance of traditional algorithms including ANR [19], GR [19] and A+ [20], as well as DL-based techniques including SRCNN [4], VDSR [11], DRCN [12], LapSRN [13], SESR [3] and our proposed RBAM. Table 2 summarizes the quantitative comparisons in terms of peak signal-to-noise-ratio (PSNR-SEM), structural similarity (SSIM), and inference time at ×2, and ×4 SR. From the table, one can see that most DL-based methods consistently outperform traditional SR algorithms in PSNR and SSIM metrics. Particularly, RBAM significantly outperforms the mean PSNR over all datasets by 0.18 dB and 0.13 dB for ×2 and ×4 SR, respectively. Furthermore, RBAM is a practical compromise between inference time, and generalization. Our results show a moderate quantitative increase in PSNR score and a considerable increase in qualitative performance - this is similar to previous

Table 2. Quantitative results of SR models at ×2 and ×4 factors. **Bold** indicates the best result. ❖ and † denote traditional and DL-based methods, respectively. PSNR scores are reported with the standard error of the mean (SEM) for each method.

Methods	CLE100		CLE200		CLE1000		time
Scale ×2	PSNR	SSIM	PSNR	SSIM	PSNR	SSIM	
Bicubic	33.69±0.06	0.8693	35.53±0.01	0.9029	34.45±0.01	0.8920	0.02
A+❖ [20]	34.22±0.07	0.8928	36.14±0.01	0.9218	35.04±0.01	0.9114	6.72
ANR❖ [19]	36.44±0.13	0.9226	39.10±0.01	0.9559	37.64±0.01	0.9559	6.07
GR❖ [19]	36.56±0.13	0.9243	39.26±0.01	0.9579	37.79±0.01	0.9448	4.47
SRCNN† [4]	35.75±0.11	0.9181	38.25±0.01	0.9494	36.87±0.01	0.9380	0.06
VDSR† [11]	36.72±0.13	0.9276	39.31±0.01	0.9578	37.89±0.01	0.9462	0.25
DRCN† [12]	36.65±0.13	0.9257	39.29±0.01	0.9575	37.83±0.01	0.9452	0.48
LapSRN† [13]	36.71±0.13	0.9264	39.25±0.01	0.9583	37.91±0.01	0.9462	0.07
SESR† [3]	36.76±0.13	0.9282	39.36±0.01	0.9583	37.91±0.01	0.9462	0.27
RBAM (Ours)†	**36.91±0.12**	**0.9321**	**39.45±0.01**	**0.9590**	**38.22±0.01**	**0.9501**	0.18
Scale ×4							
Bicubic	31.29±0.04	0.6673	32.45±0.01	0.7318	31.78±0.01	0.7278	0.02
A+ [20]	31.57±0.04	0.7042	32.76±0.01	0.7607	32.06±0.01	0.7517	3.03
ANR [19]	31.68±0.04	0.7160	32.93±0.01	0.7736	32.23±0.01	0.7671	2.88
GR [19]	31.70±0.04	0.7201	32.95±0.01	0.7736	32.25±0.01	0.7703	2.31
SRCNN [4]	31.59±0.04	0.7073	32.76±0.01	0.7617	32.07±0.01	0.7566	0.06
VDSR [11]	31.66±0.04	0.7144	32.86±0.01	0.7804	32.16±0.01	0.7635	0.25
DRCN [12]	31.70±0.04	0.7214	32.92±0.01	0.7750	32.21±0.01	0.7635	0.48
LapSRN [13]	31.68±0.04	0.7190	32.76±0.01	0.7617	32.29±0.01	0.7737	0.08
SESR [3]	31.76±0.04	0.7249	32.99±0.01	0.7804	32.29±0.01	0.7737	0.33
RBAM (Ours)	**31.84±0.04**	**0.7315**	**33.11±0.01**	**0.7852**	**32.47±0.01**	**0.7874**	0.07

works in single image super resolution [21]. Figure 3 shows selected image patches from each dataset for qualitative assessment. RBAM can delicately restore high-frequency cues, such as granular textures and sudden changes in grayscale pixel intensity. This manifests qualitatively in the form of improved restoration of high frequency details such as cell membranes (CLE200, CLE1000 examples) and intracellular spaces (CLE100 example).

Motivation for Bilinear Pooling. We combine 1st-order and 2nd-order pooling to recalibrate learned features based on channels that activate often or correspond to feature rich inputs, respectively. Channels that activate often are likely responding to common, low frequency image features. Conversely, channels that are highly correlated may be responding to feature rich instances in the image space that activate multiple filters simultaneously. High frequency features tend to be complex, and not as common semantically compared to low frequency image details. Therefore, channels that learn complex image features may not be emphasized by first order pooling operations alone. Combining first and second order pooling in an attention module assures that hard working channels are rewarded without diminishing the optimization of channels that learn complex features in the low to high resolution image mapping space.

4   Conclusion

We proposed the first network that simultaneously leverages both first and second order statistics for pooling in spatial and channel-wise attention mechanisms, resulting in a lightweight and fast model that restores high frequency image details. We compared our proposed model with various traditional and DL-based SR techniques on three CLE datasets in terms of image quality assessment metrics and inference time. Our RBAM network outperforms existing lightweight methods across different datasets, downsampling factors, and SR performance evaluation criteria. Experimental results also highlight the potential applicability of inexpensive software-based post-processing SR modules that improve degraded images in miniaturized CLE devices in real-time.

Acknowledgments. Thanks to the NVIDIA Corporation for the donation of Titan X GPUs used in this research and to the Collaborative Health Research Projects (CHRP) for funding.

References

1. Ahn, N., et al.: Fast, accurate, and lightweight super-resolution with cascading residual network. In: ECCV (2018)
2. Bray, F., et al.: Global cancer statistics 2018: GLOBOCAN estimates of incidence and mortality worldwide for 36 cancers in 185 countries. CA Cancer J. Clin. **68**(6), 394–424 (2018)
3. Cheng, X., et al.: Sesr: single image super resolution with recursive squeeze and excitation networks. In: IEEE ICPR, pp. 147–152 (2018)
4. Dong, C., et al.: Image super-resolution using deep convolutional networks. IEEE TPAMI **38**(2), 295–307 (2016)
5. Gao, Z., et al. Global second-order pooling neural networks. arXiv:1811.12006 (2018)
6. Grisan, E., et al.: 239 computer aided diagnosis of barrett's esophagus using confocal laser endomicroscopy: preliminary data. Gastrointest. Endosc. **75**(4), AB126 (2012)
7. Hu, J., et al.: Squeeze-and-excitation networks. In: IEEE CVPR (2018)
8. Huang, J., et al.: Single image super-resolution from transformed self-exemplars. In: IEEE CVPR, pp. 5197–5206 (2015)
9. Izadi, S., Moriarty, K.P., Hamarneh, G.: Can deep learning relax endomicroscopy hardware miniaturization requirements? In: Frangi, A.F., Schnabel, J.A., Davatzikos, C., Alberola-López, C., Fichtinger, G. (eds.) MICCAI 2018. LNCS, vol. 11070, pp. 57–64. Springer, Cham (2018). https://doi.org/10.1007/978-3-030-00928-1_7
10. Kiesslich, R., et al.: Confocal laser endoscopy for diagnosing intraepithelial neoplasias and colorectal cancer in vivo. Gastroenterology **127**(3), 706–713 (2004)

11. Kim, J., et al.: Accurate image super-resolution using very deep convolutional networks. In: IEEE CVPR, pp. 1646–1654 (2016)
12. Kim, J., et al.: Deeply-recursive convolutional network for image super-resolution. In: IEEE CVPR, pp. 1637–1645 (2016)
13. Lai, W., et al.: Fast and Accurate Image Super-resolution with Deep Laplacian Pyramid Networks. In: Ieee Tpami, p. 1 (2018)
14. Leong, R.W., et al.: In vivo confocal endomicroscopy in the diagnosis and evaluation of celiac disease. Gastroenterology $135(6)$, 1870–1876 (2008)
15. Ravì, D., et al.: Effective deep learning training for single-image super-resolution in endomicroscopy exploiting video-registration-based reconstruction. Int. J. Comput. Assist. Radiol. Surg. 13, 917–924 (2018)
16. Ravì, D., et al.: Adversarial training with cycle consistency for unsupervised super-resolution in endomicroscopy. Med. Image Anal. 53, 123–131 (2019)
17. Shi, W., et al.: Real-time single image and video super-resolution using an efficient sub-pixel convolutional neural network. In: IEEE Conference on Computer Vision and Pattern Recognition (CVPR), pp. 1874–1883 (2016)
18. Ştefănescu, D., et al.: Computer aided diagnosis for confocal laser endomicroscopy in advanced colorectal adenocarcinoma. PloS ONE $11(5)$, e0154863 (2016)
19. Timofte, R., et al.: Anchored neighborhood regression for fast example-based super-resolution. In: IEEE ICCV, pp. 1920–1927 (2013)
20. Timofte, R., et al.: A+: adjusted anchored neighborhood regression for fast super-resolution. In: ACCV, pp. 111–126 (2015)
21. Yang, W., Zhang, X., Tian, Y., Wang, W., Xue, J.-H., Liao, Q.: Deep learning for single image super-resolution: a brief review. IEEE Trans. Multimedia $\mathbf{PP}(99)$, 1 (2019)

TPSDicyc: Improved Deformation Invariant Cross-domain Medical Image Synthesis

Chengjia Wang[1(✉)], Giorgos Papanastasiou[2], Sotirios Tsaftaris[3], Guang Yang[4], Calum Gray[2], David Newby[1], Gillian Macnaught[2], and Tom MacGillivray[2]

[1] BHF Centre for Cardiovascular Science, University of Edinburgh,
Edinburgh EH16 4TJ, UK
chengjia.wang@ed.ac.uk
[2] Edinburgh Imaging Facility QMRI, University of Edinburgh,
Edinburgh EH16 4TJ, UK
[3] Institute for Digital Communications, School of Engineering,
University of Edinburgh, West Mains Rd, Edinburgh EH9 3FB, UK
[4] National Heart and Lung Institute, Imperial College London,
London SW3 6LY, UK

Abstract. Cycle-consistent generative adversarial network (CycleGAN) has been widely used for cross-domain medical image systhesis tasks particularly due to its ability to deal with unpaired data. However, most CycleGAN-based synthesis methods can not achieve good alignment between the synthesized images and data from the source domain, even with additional image alignment losses. This is because the CycleGAN generator network can encode the relative deformations and noises associated to different domains. This can be detrimental for the downstream applications that rely on the synthesized images, such as generating pseudo-CT for PET-MR attenuation correction. In this paper, we present a deformation invariant model based on the deformation-invariant CycleGAN (DicycleGAN) architecture and the spatial transformation network (STN) using thin-plate-spline (TPS). The proposed method can be trained with unpaired and unaligned data, and generate synthesised images aligned with the source data. Robustness to the presence of relative deformations between data from the source and target domain has been evaluated through experiments on multi-sequence brain MR data and multi-modality abdominal CT and MR data. Experiment results demonstrated that our method can achieve better alignment between the source and target data while maintaining superior image quality of signal compared to several state-of-the-art CycleGAN-based methods.

1 Introduction

Cross-domain image synthesis is gaining popularity in a wide range of clinical applications to enable multi-modality synthesis without acquiring data from multiple modalities. However, the vast majority of cross-modality synthesis methods are solely evaluated on brain image data due to the low geometric variance. Otherwise, performance of the synthesis methods often rely on a registration-based

© Springer Nature Switzerland AG 2019
F. Knoll et al. (Eds.): MLMIR 2019, LNCS 11905, pp. 245–254, 2019.
https://doi.org/10.1007/978-3-030-33843-5_23

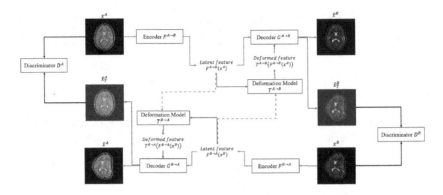

Fig. 1. Architecture of TPSDicyc

preprocessing step. Previous studies have shown that CycleGAN [1] achieves high synthesis quality on unpaired data, but it has also been observed that Cycle-GAN may reproduce the "domain-specific deformations" of the data [2,3]. A common strategy to address this issue is to leverage image similarity metrics in the CycleGAN loss function [4,5], but this introduces the trade-off between good quality of signal and good data alignment. The deformation-invariant CycleGAN (DicycleGAN) model [3] has been proposed recently achieved state-of-the-art synthesis performances with better alignment of data. This method uses two sets of parameters to encode translatable appearance features and relative spatial deformations between the training images using the deformable convolution (DC) operation [6]. However, DicycleGAN models a relatively consistent local deformation between the source and target data. This limits the generalizability of the model on multiple scanners, and requires the subjects in similar poses when being imaged. Otherwise the learning process can be unstable and slow to converge as shown in our experiments.

In this paper, we present an alternative framework of DicycleGAN based on thin-plate-spline (TPS) named as TPSDicyc. Compared to DicycleGAN, which models combines the "deformation" and "image translation" parameters into one network, TPSDicyc uses a separated spatial transformation network (STN) to learn the relative deformation between the source and target data. Figure 1 presents the TPSDicyc framework and its subnetworks. Figure 2 displays the architecture of TPSDicyc generator network. We evaluated the proposed method using both publicly available multi-sequence brain MR data and multi-modality abdominal data. Compared to the selected state-of-the-art baseline methods, TPSDicyc displayed better ability to handle disparate imaging domains and to generate synthesized images aligned with the source data.

2   Previous Works

CycleGAN was first applied to cross-domain medical image synthesis in [7] for co-synthesis of CT and MR brain data. Alignment between the synthesised data

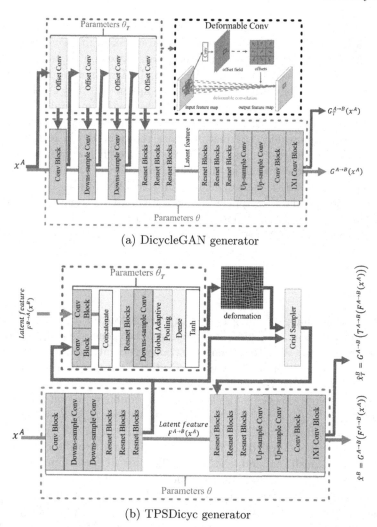

(a) DicycleGAN generator

(b) TPSDicyc generator

Fig. 2. Structures and parameters of generators in DicycleGAN and our TPSDicyc: DicycleGAN uses deformable convolutional layers to model the relative deformation between the source and target data; in TPSDicyc the deformation is learned by a separated Thin-plate-spline based spatial transformation network. (Color figure online)

and the source data can be improved by regularizing the problem through multi-task training, for example, using segmentation masks [2], and by co-registration [7]. However, these models have an extra cost of manual annotations for segmentation or registration ground truths. A currently popular strategy is to integrate image similarity measures into the CycleGAN loss so that the geometric correspondences between data from different domains can be improved. For example, [5] introduces a structure-consistency loss based on the modality independent neighborhood descriptor (MIND) [8]. It has been shown that this structure-constrained CycleGAN can be trained with unregistered multi-modal MR and CT brain data. A similar gradient-consistency loss, based on the normalised gradient cross correlation (GCC), is introduced in [4]. This method has been evaluated using unpaired but pre-registered, multi-modal MR and CT hip images. However, as discussed in [3], there is a conflict between the image similarity based losses and the CycleGAN discriminative loss. Because the synthesized data in which the "domain-specific doformations" are reproduced will lead to a lower adversarial loss (of the discriminator in GANs) but higher alignment loss. As a result, the synthesized data can not be well aligned to the source data and show a good quality of signal at the same time. DicycleGAN [3] uses DC parameters to decouple the translatable appearance features and the relative deformation between the source and target data, thus introduces a possible solution of the conflicts in the CycleGAN losses. However, DC layers can only learn relative consistent and local deformations.

3   Method

Assuming that we have n^A images $x^A \in \mathcal{X}^A$ from domain $\mathcal{X}^A$, and n^B images $x^B \in \mathcal{X}^B$ from domain $\mathcal{X}^B$, synthesis is performed to generate images of domain $\mathcal{X}^B$ using images from $\mathcal{X}^A$. To this end, we train a generator (which consists of an encoder $F^{A \to B}$, a decoder $G^{A \to B}$ and a STN $T^{A \to B}$) and a discriminator D^B in the min-max game of the GAN loss $\mathcal{L}_{GAN}\left(F^{A \to B}, T^{A \to B}, G^{A \to B}, D^B, \mathcal{X}^A, \mathcal{X}^B\right)$. We let $\mathcal{L}_{GAN}^{A \to B}$ denote this GAN loss for short and simple representation. Accordingly, $G^{B \to A}$, $F^{B \to A}$, $T^{B \to A}$, D^A, and the GAN loss $\mathcal{L}_{GAN}^{B \to A}$ are defined. A CycleGAN-based framework consists of two symmetric sets of generators act as mapping functions applied to a source domain, and two discriminators D^B and D^A to distinguish real and synthesized data for a target domain. The *cycle consistency* loss $\mathcal{L}_{cyc}^{A,B}$, is used to keep the cycle-consistency between the two sets of networks. This gives CycleGAN the ability to deal with unpaired data. Then the loss of the whole CycleGAN framework $\mathcal{L}_{CycleGAN}$ is $\mathcal{L}_{CycleGAN} = \mathcal{L}_{GAN}^{A \to B} + \mathcal{L}_{GAN}^{B \to A} + \lambda_{cyc}\mathcal{L}_{cyc}^{A,B}$. Presently proposed CycleGAN-based methods add an image alignment term $\mathcal{L}_{align}^{A,B}$ to $\mathcal{L}_{CycleGAN}$ which becomes $\mathcal{L}_{CycleGAN,align} = \mathcal{L}_{CycleGAN} + \lambda_{align}\mathcal{L}_{align}^{A,B} = \mathcal{L}_{GAN}^{A \to B} + \mathcal{L}_{GAN}^{B \to A} + \lambda_{cyc}\mathcal{L}_{cyc}^{A,B} + \lambda_{align}\mathcal{L}_{align}^{A,B}$, where λ_{align} is the weight used to balance $\mathcal{L}_{align}^{A,B}$ and $\mathcal{L}_{CycleGAN}$.

3.1   Architecture of Generator

As shown in Fig. 2, the generator network consists of an encoder, an decoder and a STN. The encoder maps the input data into a latent feature space, and the decoder generates the synthesized image based on the latent features. The relative deformation between the source and target domains are learned by a subset of parameters θ_T which are only used in the training process. In [3], θ_T is the trainable parameters in the DC layers. In this work, we introduced a new spatial transformation sub-network T for this purpose. As shown in Fig. 2, the transformation sub-network take latent features extracted from the source and target data, and produces a displacement field of a keypoint grid. This displacement between keypoints are then applied to the source latent features using thin-plate-spline (TPS) interpolation. In this case, θ_T represents the parameters in this spatial transformation subnet, and θ the rest parameters in G. To generate synthesised images for domain $\mathcal{X}^B$, each input image x^A generates two output images through two separated forward passes: deformed output image $\hat{x}_T^B = G^{A \to B} \left(T^{A \to B} \left(F^{A \to B} \left(x^A \right) \right) \right)$ and undeformed image $\hat{x}^B = G^{A \to B} \left(F^{A \to B} \left(x^A \right) \right)$. $\hat{x}_T^B$ generated by passing the latent feature $F(x)$ through T (shown by the red arrows in Fig. 2) is expected to be identical to x^B. $\hat{x}^B$ is expected to be aligned with x^A.

3.2 Loss and Training

Similar to DicycleGAN, TPSDicyc loss functions include the traditional GAN loss, the cycle-consistency loss used in the original CycleGAN, an image alignment loss and an additional deformation invariant cycle consistency loss.

For the GAN loss $\mathcal{L}_{GAN}^{A \to B}$, $F^{A \to B}$, $G^{A \to B}$ and $T^{A \to B}$ are trained to minimize $\left(D^B \left(\hat{x}_T^B \right) - 1 \right)^2$ and D^B is trained to minimize $\left(D^B (x^B) - 1 \right)^2 + D^B \left(\hat{x}_T^B \right)^2$. The same formulation is used to calculate $\mathcal{L}_{GAN}^{B \to A}$ defined on the "$B \to A$" networks. Note that the GAN loss is calculated based on the deformed outputs. As the undeformed outputs of generators are expected to be aligned with the input images, an image alignment loss based on normalized mutual information (NMI) is defined as:

$$\mathcal{L}_{align}^{A,B} = 2 - NMI \left(x^A, \hat{x}^B \right) - NMI \left(x^B, \hat{x}^A \right). \tag{1}$$

Essentially this image alignment loss can be adopted with any similarity measure suitable for image registration, such as normalized mutual information (NMI) [9], normalised GCC used in [4], or MIND in [5] and [8].

Secondly, the cycle-consistency loss plays a critical role for the outstanding performance of CycleGAN. In this work, both the undeformed and deformed version of synthesized data should be cycle-consistent to encode optimal representations. This results in two cycle-consistency losses. The undeformed cycle consistency loss is defined as:

$$\mathcal{L}_{cyc}^{A,B} = \| G^{B \to A} \left(F^{B \to A} \left(\hat{x}^B \right) \right) - x^A \|_1 + \| G^{A \to B} \left(F^{A \to B} \left(\hat{x}^A \right) \right) - x^B \|_1, \tag{2}$$

and the deformation-invariant cycle consistency loss is:

$$\begin{aligned} \mathcal{L}_{dicyc}^{A,B} = & \| G^{B \to A} \left(T^{B \to A} \left(F^{B \to A} \left(\hat{x}^B \right) \right) \right) - x^A \|_1 \\ & + \| G^{A \to B} \left(T^{A \to B} \left(F^{A \to B} \left(\hat{x}^A \right) \right) \right) - x^B \|_1. \end{aligned} \tag{3}$$

The complete TPSDicyc loss is then defined as:

$$\mathcal{L}_{TPSDicyc} = \mathcal{L}_{GAN}^{A \to B} + \mathcal{L}_{GAN}^{B \to A} + \lambda_{align} \mathcal{L}_{align}^{A,B} + \lambda_{cyc} \mathcal{L}_{cyc}^{A,B} + \lambda_{dicyc} \mathcal{L}_{dicyc}^{A,B}. \tag{4}$$

In this work, we set $\lambda_{cyc} = \lambda_{dicyc} = 10$ and $\lambda_{align} = 0.9$. The models were trained with Adam optimizer [10] with a fixed learning rate of 0.0002 for the first 100 epochs, followed by 100 epochs with linearly decreasing learning rate. Here we apply a simple early stop strategy: in the first 100 epochs, when $\mathcal{L}_{TPSDicyc}$ stops decreasing for 10 epochs, the training will move to the learning rate decaying stage; similarly, this tolerance is set to 20 epochs in the second 100 epochs.

4   Experiments

IXI dataset: The Information eXtraction from Images (IXI) dataset[1] provides co-registered multi-sequence MR images collected from multiple sites. We used 66 pairs of proton density (PD-) and T2-weighted volumes for T2→PD synthesis experiment, each volume has 116 to 130 slices. We use 38 pairs for training and 28 pairs for evaluation of synthesis results. Our image generators take 2D axial-plane slices of the volumes as inputs. All volumes were resampled to a resolution of $1.8 \times 1.8 \times 1.8 \, \text{mm}^3/\text{voxel}$, then cropped to a size of 128×128 pixels. All the images are bias field corrected and normalized with their mean and standard deviation. We applied a simulated deformation to all T2-weighted images. Synthesis experiments were then performed between the undeformed PD-weighted data and the deformed T2-weighted data. When using deformed T2-weighted images, the ground truths of synthesized PD-weighted data were generated by applying the same nonlinear deformation to the source PD-weighted images.

Private Abdominal Data. We used a dataset containing 40 multi-modality abdominal T2*-weighted and CT images collected from 20 patients with abdominal aortic aneurysm (AAA). All images are resampled to a resolution of $1.56 \times 1.56 \times 5 \, \text{mm}^3/\text{voxel}$, and the axial-plane slices trimmed to 192×192 pixels. Because of the "domain-specific deformations", registration based ground truths as in the IXI dataset are not available. However, because several organs, such as aorta and spine, are relatively rigid compared to other surrounding soft tissues such as lower gastrointestinal tract organs, these objects can be affinely registered for evaluation of synthesis. For each volume in the MA^3RS dataset, the anatomy of the aorta were manually segmented for each volume (as described in [11]). The multi-modality data acquired from the same patient were affinely registered so that the segmented aorta in both data are well aligned. The manual registration and segmentation were performed by 4 clinical researchers. Signal of the synthesized images were evaluated within the segmentation of aorta.

4.1   Evaluation Metrics

To be consistent with the baseline methods, we use three metrics to evaluate performance on cross-domain image synthesis: mean squared error (MSE), peak signal-to-noise ratio (PSNR) and structural similarity index (SSIM) as typically

---

[1]   http://brain-development.org/ixi-dataset/.

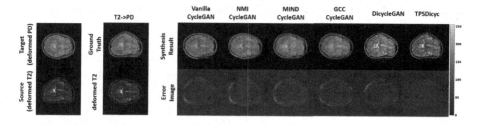

Fig. 3. Visualized PD→T2 synthesis results of the IXI dataset: an arbitrary deformation was applied to the T2 weighted images.

used by other CycleGAN based methods. Given a volume x^A and a target volume x^B, the MSE is computed as: $\frac{1}{N}\sum_1^N \left(x^B - \hat{x}^B\right)^2$, where N is number of voxels in the volume. PSNR is calculated as: $10\log_{10}\frac{\max_B^2}{MSE}$. SSIM is computed as: $\frac{(2\mu_A\mu_B+c_1)(2\delta_{AB}+c2)}{(\mu_A^2+\mu_B^2+c_1)(\delta_A^2+\delta+B^2+c2)}$, where μ and δ^2 are mean and variance of a volume, and δ_{AB} is the covariance between x^A and x^B. c_1 and c_2 are two variables to stabilize the division with weak denominator [12]. Larger PSNR and SSIM, or smaller MSE, indicate a better performance of a synthesis algorithm. To test the statistical significance of results, we perform paired t-test between the TPSDicyc and the DicycleGAN baseline. Differences between performances are considered to be statistically significant when the p−value is less than 0.05.

4.2 Results and Discussion

IXI Dataset. The quantitative results is shown in Table 1. Vanilla CycleGAN trained on paired and registered images (without simulated deformation) a theoretical upper-bound performance with PSNR > 24.3, SSIM > 0.817 and MSE ≤ 0.036. Trained with unpaired data suffering from simulated deformation, the vanilla CycleGAN gave a lower-bound baseline of performances. With additive image alignment losses, GCC-CycleGAN [4] and MIND-CycleGAN [5] methods lead to tiny improvements in terms of PSNR. However, because these two models are still affected by the simulated "domain-specific deformation", their performances were still comparable to vanilla CycleGAN. In contrast, the proposed TPSDicyc model lead to results significantly closer to the upper-bound baseline.

Alignment between source and target data can be observed in the example shown in Fig. 3. It can be seen that the vanilla CycleGAN model exactly reproduced the simulated deformation. The GCC-CycleGAN and MIND-CycleGAN, although can reduce the misalignment effect, the synthesized and source data are still not well aligned. Furthermore, the synthesis results generated by the three CycleGAN-based models are blurry and showed visible artifacts. In contrast, our TPSDicyc model achieved best data alignment.

Abdominal data: Table 2 shows the quantitative assessments of the four compared models based on the same metrics used for the IXI data. The vanilla CycleGAN had slightly better performances compared the GCC- and MIND-CycleGAN models. Our method lead to over 20% performance gains in terms

Table 1. Synthesis results of IXI dataset using deformed T2 images.

	Method	MSE	PSNR	SSIM
	Cycle [7]	0.055 (0.22)	20.80 (2.87)	0.708 (0.19)
T2	GCC-Cycle [4]	0.054 (0.22)	21.04 (3.83)	0.719 (0.19)
↓	MIND-Cycle [5]	0.054 (0.21)	20.82 (2.61)	0.703 (0.19)
PD	DicycleGAN	0.045 (0.21)	22.52 (2.91)	0.790 (0.18)
	TPSDicyc	**0.044 (0.23)**	**22.72 (2.86)**	**0.796 (0.16)**
	Cycle (aligned)	0.037 (0.22)	24.77 (3.30)	0.856 (0.17)

Source (T2*)   Target (CT)   TPSDicyc CT   DicycleGAN CT   TPSDicyc CT   DiCycleGAN CT

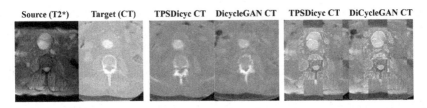

Fig. 4. Visualization of cross-modality synthesis results obtained with our MA^3RS dataset. A example data from both the CT and T2* domain are shown on the left. A checkerboard view combing the source and synthesized data is shown on the right. Alignment between the source and the synthesized data can then be assessed by looking at the anatomy of aorta and spine, as well as the lower contour of the patient body.

Table 2. T2*→CT synthesis results using private dataset.

T2* → CT			
Model	MSE	PSNR	SSIM
Cycle [7]	0.009 (0.004)	20.57 (2.12)	0.675 (0.06)
GCC-Cycle [4]	0.012 (0.006)	20.25 (2.35)	0.602 (0.08)
MIND-Cycle [5]	0.010 (0.004)	21.21 (2.04)	0.660 (0.07)
DicycleGAN	0.008 (0.004)	22.01 (2.40)	0.694 (0.07)
TPSDicyc	**0.008 (0.004)**	**22.29 (2.26)**	**0.706 (0.06)**

of MSE and SSIM, and also achieved better performance compared to Dicycle-GAN. Except for SSD, p-value of the paired t-test between DicycleGAN and our method are less than 0.05. Figure 4 provides a checkerboard visualization combining the source image and synthesized data generated by the DicycleGAN and our TPSDicyc. Objects such as spine and aorta in the source and target data can only be affinely registered independently. Both DicycleGAN and TPS-Dicyc model produce synthesized images where these objects are simultaneously aligned in the source and target data. DicycleGAN achieved better alignment of the outer contour of image subject while TPSDicyc show better alignment for spine and aorta.

5   Conclusion

In this paper, we propose the Tpsdicyc model to address the issue of "domain-specific deformation". Different from the recently proposed DicycleGAN model, we integrate a TPS-based spatial transformation sub-network in the CycleGAN model and train the model with associated deformation-invariant cycle consistency loss and NMI-based alignment loss function. Compared to the DC layers in DicycleGAN, this new architecture allows to model global deformations. Our Tpsdicyc method can achieve good alignment between the source and synthesized data, and outperformed the DicycleGAN, as well as state-of-the-art CycleGAN-based models in experiments performed on multi-sequence MR data and multi-modality abdominal data.

Acknowledgments. This work is funded by British Heart Fundation (no. RG/16/10/32375). S.A. Tsaftaris and G. Papanastasiou acknowledge support from the EPSRC Grant (EP/P022928/1). Support from NHS Lothian R&D, and Edinburgh Imaging and the Edinburgh Clinical Research Facility at the University of Edinburgh is gratefully acknowledged.

References

1. Zhu, J.Y., Park, T., Isola, P., Efros, A.A.: Unpaired image-to-image translation using cycle-consistent adversarial networks. arXiv preprint arXiv:1703.10593 (2017)
2. Chartsias, A., Joyce, T., Dharmakumar, R., Tsaftaris, S.A.: Adversarial image synthesis for unpaired multi-modal cardiac data. In: Tsaftaris, S.A., Gooya, A., Frangi, A.F., Prince, J.L. (eds.) SASHIMI 2017. LNCS, vol. 10557, pp. 3–13. Springer, Cham (2017). https://doi.org/10.1007/978-3-319-68127-6_1
3. Wang, C., Macnaught, G., Papanastasiou, G., MacGillivray, T., Newby, D.: Unsupervised learning for cross-domain medical image synthesis using deformation invariant cycle consistency networks. In: Gooya, A., Goksel, O., Oguz, I., Burgos, N. (eds.) SASHIMI 2018. LNCS, vol. 11037, pp. 52–60. Springer, Cham (2018). https://doi.org/10.1007/978-3-030-00536-8_6
4. Hiasa, Y., et al.: Cross-modality image synthesis from unpaired data using cyclegan: effects of gradient consistency loss and training data size. arXiv preprint arXiv:1803.06629 (2018)
5. Yang, H., et al.: Unpaired brain MR-to-CT synthesis using a structure-constrained cyclegan. In: Stoyanov, D., et al. (eds.) DLMIA/ML-CDS -2018. LNCS, vol. 11045, pp. 174–182. Springer, Cham (2018). https://doi.org/10.1007/978-3-030-00889-5_20
6. Dai, J., Qi, H., Xiong, Y., Li, Y., Zhang, G., Hu, H., Wei, Y.: Deformable convolutional networks. CoRR, abs/1703.06211, vol. 1, no. 2 (2017). 3
7. Wolterink, J.M., Dinkla, A.M., Savenije, M.H.F., Seevinck, P.R., van den Berg, C.A.T., Išgum, I.: Deep MR to CT synthesis using unpaired data. In: Tsaftaris, S.A., Gooya, A., Frangi, A.F., Prince, J.L. (eds.) SASHIMI 2017. LNCS, vol. 10557, pp. 14–23. Springer, Cham (2017). https://doi.org/10.1007/978-3-319-68127-6_2
8. Heinrich, M.P., et al.: Mind: modality independent neighbourhood descriptor for multi-modal deformable registration. Med. Image Anal. **16**(7), 1423–1435 (2012)

9. Vinh, N.X., Epps, J., Bailey, J.: Information theoretic measures for clusterings comparison: variants, properties, normalization and correction for chance. J. Mach. Learn. Res. **11**(Oct), 2837–2854 (2010)
10. Kingma, D.P., Ba, J.: Adam: a method for stochastic optimization. arXiv preprint arXiv:1412.6980 (2014)
11. Papanastasiou, G., et al.: Multidimensional assessments of abdominal aortic aneurysms by magnetic resonance against ultrasound diameter measurements. In: Valdés Hernández, M., González-Castro, V. (eds.) MIUA 2017. CCIS, vol. 723, pp. 133–143. Springer, Cham (2017). https://doi.org/10.1007/978-3-319-60964-5_12
12. Hore, A., Ziou, D.: Image quality metrics: PSNR vs. SSIM. In: 2010 20th international conference on Pattern recognition (ICPR), pp. 2366–2369. IEEE (2010)

PredictUS: A Method to Extend the Resolution-Precision Trade-Off in Quantitative Ultrasound Image Reconstruction

Farah Deeba[1]([✉])[iD] and Robert Rohling[1,2]

[1] Department of Electrical and Computer Engineering, The University of British Columbia, Vancouver, BC, Canada
{farahdeeba,rohling}@ece.ubc.ca
[2] Department of Mechanical Engineering, The University of British Columbia, Vancouver, BC, Canada

Abstract. We present PredictUS, a novel Quantitative Ultrasound (QUS) parameter estimation technique with improved resolution and precision using augmented ultrasound data. The ultrasound data is generated using a sequence-to-sequence convolutional neural network based on WaveNet. The spectral-based QUS techniques are limited by the well-studied trade-off between the precision of the estimated QUS parameters and the window size used in estimation, limiting the practical utility of the QUS techniques. In this paper, we present a method to increase the window size by predicting the next data points of a given window. The method provides better estimates of local tissue properties with high resolution by virtually extending the property to a larger region. Our proof-of-concept study based on attenuation coefficient estimate (ACE), an important QUS parameter, attains a resolution reduction up to 50% while maintaining comparable estimation precision. This result shows the promise to extend the precision-resolution trade-off, which, in turn, would have implications in small lesion detection or heterogeneous tissue characterization.

Keywords: Quantitative Ultrasound · Attenuation coefficient estimate · WaveNet · Sequence-to-sequence neural network

1 Introduction

Quantitative Ultrasound (QUS) Imaging has introduced a paradigm shift in the field of biomedical imaging. Extending beyond qualitative B-mode ultrasound imaging, QUS presents clinically significant parametric images, which are descriptive of underlying tissue microstructure. Recent studies show that QUS potentially provides effective, non-invasive, and system independent biomarkers for non-alcoholic fatty liver disease (NAFLD) detection and monitoring, cervical

© Springer Nature Switzerland AG 2019
F. Knoll et al. (Eds.): MLMIR 2019, LNCS 11905, pp. 255–264, 2019.
https://doi.org/10.1007/978-3-030-33843-5_24

ripening detection, placenta characterization, and breast lesion characterization [1–3].

QUS extracts acoustic scattering and attenuating properties using algorithms based on estimates of power spectra of ultrasound radiofrequency (RF) signal backscattered from the interrogated tissue. The spectral based QUS techniques allow the normalization of the backscattered RF signal and thus filter out the system-dependent factors such as focusing, diffraction and transducer electromechanical response [1,4]. Unfortunately, the power spectral estimation, typically obtained from FFT based periodogram of windowed RF signal, imposes a fundamental trade-off on QUS between image resolution and estimation precision. Smaller windows provide high spatial resolution, a desirable property for many imaging applications such as characterization of thin (e.g. human skin) or heterogeneous (e.g. placenta) tissue. However, smaller windows yield noisy and inaccurate power spectra estimates due to limited spectral resolution and spatial variation noise inherent in ultrasonic scattering. Larger windows improve accuracy and precision of power spectra estimates and therefore the estimation of QUS parameters, with an expense of reduced spatial resolution [5,6]. One study found that the trade-off between spatial resolution and the variance of QUS parameter is optimized with a window size of 10 independent scanlines laterally and 10 times the wavelength axially [6].

To expand the precision-resolution trade-off, different modifications of periodogram have been investigated. Welch method was found to yield the most accurate and precise spectral estimate with reasonable computational cost [6]. Alternatively, autoregressive (AR) techniques have been reported to exceed the performance of FFT based periodogram, especially for smaller windows [7]. However, AR techniques show degraded performance with increasing depth in higher attenuating media due to violation of the stationarity assumption. More recently, deep neural networks such as WaveNet [8], have significantly improved state-of-the-art performance in fields of forecasting non-stationary and non-linear processes, such as speech and financial time-series [8,9].

In this work, we present PredictUS, a spectral based QUS technique based on US RF signal prediction using a sequence-to-sequence convolutional neural network (CNN) modelled with a WaveNet inspired architecture. Given a small windowed RF signal, this method predicts the next data points, resulting in a larger window. Therefore, the method yields better estimates of power spectra as well as can characterize local tissue properties with high resolution by essentially extending the property to a larger region. We demonstrate the applicability of PredictUs for improved measurement of attenuation coefficient estimate (ACE), a QUS parameter.

2   Method

The first signal processing step in spectral-based QUS techniques is the estimation of power spectra from a limited-length RF signal window. The proposed PredictUS method adds a WaveNet inspired deep neural network before the

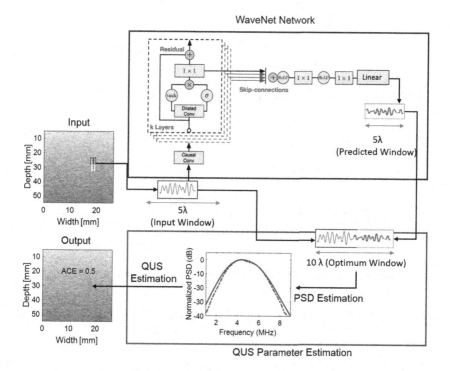

Fig. 1. Overview of the proposed PredictUS method. Blue solid line indicates the original input window, whereas the blue dash-dotted line indicates the predicted signal in the "WaveNet Network" block and the power spectrum obtained from the optimum (i.e. PredictUS) window in the "QUS Parameter Estimation" block. (Color figure online)

power spectra estimation step (Fig. 1). The network results in a larger window by sequentially predicting the next samples. The power spectra estimated from the larger RF signal windows are fed for the subsequent QUS parameter estimation.

2.1 ACE Computation

The ultrasound ACE is a measure of ultrasound amplitude dissipation due to the combined effect of scattering and absorption. ACE can be measured using the reference phantom method [4], a standard method to account for the system dependent factors.

According to this method, the RF data are acquired from both the tissue sample, s and a reference phantom, r with known properties using the same transducer and system settings. For a RF signal window centered at depth z from the transducer surface, the natural logarithm of the ratio of the power spectrum S from the sample to the reference phantom at frequency $f \in (f_1, f_k)$

can be written as [2,4]:

$$\ln \frac{S_r(f,z)}{S_s(f,z)} = -4(\alpha_s(z) - \alpha_r(z))fz + \ln \frac{B_s}{B_r}. \tag{1}$$

where α is the effective ACE for the total ultrasound propagation path z and B is the backscatter coefficient (BSC). Substituting the following variables in Eq. 1 as: $\ln \frac{S_r(f,z)}{S_s(f,z)} = Y(f,z), \alpha_r - \alpha_s = \alpha, \ln \frac{B_s}{B_r} = \beta$, we get,

$$Y(f,z) = -4\alpha(z)fz + \beta. \tag{2}$$

The above equation can be written in a matrix form: $\mathbf{y} = \mathbf{Ax} + \mathcal{N}(0, \sigma_N \mathbf{I})$, where

$$\mathbf{A} = \begin{bmatrix} 4zf_1 & 1 \\ \vdots & \vdots \\ 4zf_k & 1 \end{bmatrix}_{k \times 2}, \mathbf{y} = \begin{bmatrix} Y(f_1, z) \\ \vdots \\ Y(f_k, z) \end{bmatrix}_{k \times 1}, \mathbf{x} = \begin{bmatrix} \alpha \\ \beta \end{bmatrix}_{2 \times 1}.$$

A least square fitting method [11] can be applied to solve for $\mathbf{x} = [\alpha, \beta]$ from the noisy estimation $\mathbf{y}$ as follows:

$$\hat{\mathbf{x}} = \arg\min_{\mathbf{x}} \{\|(\mathbf{y} - \mathbf{Ax})\|_2^2\}, \tag{3}$$

with the following constraints:

$$\alpha_{min} \leq \alpha \leq \alpha_{max}, \beta_{min} \leq \beta \leq \beta_{max} \tag{4}$$

Solving Eq. 2 gives us the effective ACE (α) for the total ultrasound propagation path. The local ACE at depth z_i can be computed as: $\alpha^{local}(z_i) = \frac{\alpha(z_i)z_i - \alpha(z_{i-1})z_{i-1}}{z_i - z_{i-1}}$.

2.2 Network Architecture

We employed a sequence-to-sequence CNN using WaveNet model. First introduced by the researchers from DeepMind, Wavenet is an autoregressive model, where each predicted sample is conditioned on the previous ones [8]. One key element of WaveNet is stacked layers of 1-dimensional dilated causal convolution. Causal convolutions are used to ensure that a prediction at time step t only depends on the previous time steps, whereas the use of a dilation rate increased as a factor of 2 results in an exponentially growing receptive field with depth. For the US RF signal prediction, a receptive field is required which is large enough to capture several wavelengths. For our application, one wavelength (λ) is approximately 20 samples for US transmission frequency of 5 MHz and sampling frequency of 50 MHz. We use 14 dilated causal convolution layers with a dilation rate of factor 2 with a reset $(2^0, 2^1, ..., 2^6, 2^0, 2^1, ..., 2^6)$ and 64 filters with width of 2. As in original architecture, there are gated activation unit (combining a hyperbolic tangent and a sigmoid activation branch), residual, and skip connections (Fig. 1) in each of these layers to speed up the convergence and enable improved training for deeper models. Finally, the WaveNet output is passed through a ReLU activation followed by a linear projection.

3   Experiments and Results

3.1   Data

We used the k-Wave toolbox [10] to generate simulation US RF data. For the simulation, we use 96 element linear array transducer with 0.2 mm element pitch. The depth was set to 60 mm. Fixed focusing was used in transmission (focal depth 60 mm) and dynamic focusing was used in reception. We simulate 256 RF lines for each of the 32 different ACE values ranging from 0.1 to 1.65 dB/cm/MHz. The selected ACE range encompasses the observed ACE values in liver at different NAFLD stages and in placenta. Finally, we divide each RF line into 13 segments with 25% overlap, resulting in a training set of 106,496 RF line segments. We also created a separate test dataset of 550 examples, where 50 RF line segments are extracted for each of the 11 different ACE values (0.5–1.5 dB/cm/MHz).

3.2   Training and Testing

In the training stage, a *teacher-forcing* procedure is applied where the model performs a one-step ahead prediction. The model outputs a sequence of n steps, which is one time-step shifted version of the input sequence. Therefore, the model is trained using the correct output instead of the predicted output. We use a batch size of 64 and use Adam optimizer to minimize mean squared error loss with $\beta = 0.9$, $\beta_2 = 0.999$ and learning rate of 0.001.

In contrast to the training stage, the testing stage makes a n-step ahead prediction sequentially by feeding each prediction back into the network at the next time step.

3.3   RF Data Processing and Analysis

For ACE computation, the lateral dimension for the RF signal window was kept fixed at 5 scanlines. The axial dimension was varied from 5λ (100 samples) to 10λ (200 samples) to study the effect of PredictUS in improving the trade-off between resolution and precision. For the power spectrum computation, the Welch method has been found to yield more accurate and precise estimation compared to rectangular, Hanning, or Hamming windows [6]. Therefore, we used the Welch method to estimate the power spectrum from the RF scanlines within each RF window. According to the Welch method, each RF scanline within a window was subdivided into overlapping sections, with length equal to 67% of the original RF scanline and with 50% overlap. Each segment was then multiplied with a Hamming window. The power spectral density was obtained after averaging the periodograms obtained from the windowed segments. We considered the −20 dB bandwidth of the received power spectrum as the usable frequency range. To compute the ACE, we utilize the RF data with ACE of 0.5 dB/cm/MHz as the reference data.

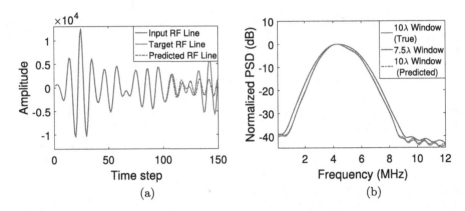

Fig. 2. (a) An example of RF line segment prediction using the proposed sequence-to-sequence CNN. (b) Comparison of power spectra estimation using original large window (10λ), original small window (7.5λ) and PredictUS large window (10λ) predicted from 7.5λ original window.

3.4   Performance Metrics

We use Mean Absolute Scaled Error (MASE) to measure the performance of sequence-to-sequence CNN to predict the larger RF window from the smaller one, where

$$MASE = \frac{\frac{1}{T}\sum_{t=1}^{T}|e_t|}{\frac{1}{T-1}\sum_{t=2}^{T}|Y_t - Y_{t-1}|} \tag{5}$$

Here, e_t is the prediction error, defined as the difference between the actual value and the predicted value and the denominator denotes the in-sample mean absolute error from the naive forecast method. A MASE value < 1 indicates a prediction performance better than the naive forecast method.

As a measure of precision of ACE, we report the standard deviation of the computed ACE as a percentage of the actual value. We also report the bias in the estimated ACE, as the difference between the estimated ACE and actual ACE presented as a percentage of the actual ACE.

3.5   Results

Performance of RF Data Prediction. We apply the trained sequence-to-sequence CNN on the test dataset. An example of RF data prediction is shown in Fig. 2a, where a 50-step (2.5λ) ahead prediction was made, given a RF line segment of 150 steps (7.5λ). We can see a precise prediction for the initial RF samples, which starts to degrade with increasing time step. We found that the power spectra estimate obtained from the PredictUS window of length 10λ (generated from 7.5λ segment) gives similar estimates as obtained from a larger window (10λ segment) and outperforms the estimates obtained from the original 7.5λ segment (Fig. 2b).

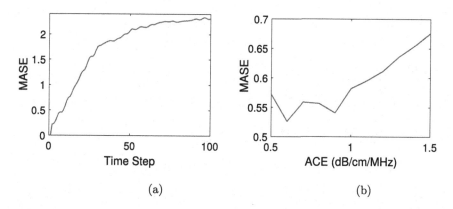

(a)   (b)

Fig. 3. RF data prediction performance: (a) for n-step ahead prediction with varying n, and (b) for varying ACE in term of mean absolute scaled error (MASE).

We analyse the performance of the network for n-step prediction for varying values of n. It was found that, on average, the network can make a 16-step prediction when $MASE < 1$. This performance is comparable to the previous work on time-series forecasting using WaveNet model, which reports MASE for a single-step ahead prediction [9]. Additionally, we investigated whether the ACE amplitude has an effect on the prediction performance. We computed MASE for a 15-step prediction for data with ACE varying from 0.5 to 1.5 dB/cm/MHz. From Fig. 3, we see that the MASE remains within a range of 0.52–0.58 for ACE $\leq$ 1 dB/cm/MHz, and after that MASE starts to degrade with increasing ACE. This result agrees with previous finding from the AR based techniques where higher attenuating medium showed inferior QUS estimation performance [7].

Performance of ACE Computation. We investigate the performance of the proposed PredictUs method in extending the resolution-precision trade-off inherent in QUS parameter estimation. According to [6], an axial dimension of 10λ has been defined to be the optimum keeping the variance and bias of QUS parameter estimation within 10%. Taken these numbers as the baseline, in this study, we examined the limit to which the trade-off can be extended by simulating three cases with increasing difficulty as follows:

1. Case I: PredictUS window of length 10λ with 2.5λ predicted using 7.5λ window;
2. Case II: PredictUS window of length 10λ with 5λ predicted using 5λ window;
3. Case III: PredictUS window of length 10λ with 7.5λ predicted using 2.5λ window;

Here, PredictUS window refers to the window obtained by concatenating the original input window and the predicted window. For each case, the precision as well as the bias in ACE computation have been compared with the original 'large'

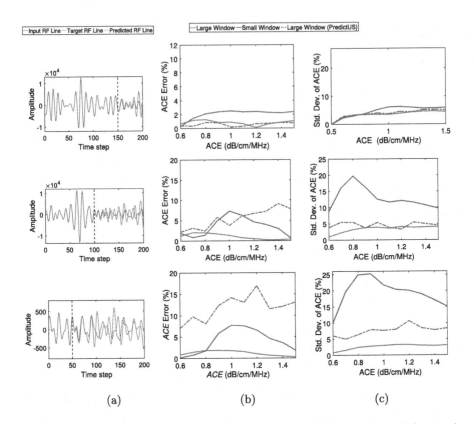

Fig. 4. PredictUS performance for case I (top), case II (middle), and case III (bottom). (a) Example RF line segments and their divisions; (b) ACE error for different ACE values; (c) Standard deviation of ACE for different ACE values.

RF window with length equal to the PredictUS window, and with the original 'small' RF window with length equal to the original data used in computing PredictUS window. The ACE computation results for these three cases have been demonstrated in Fig. 4.

For case I, the small window (7.5λ) with original data attains a precision and accuracy performance where ACE error and the standard deviation remain within 4% and 7%, respectively. The large window (10λ) with original data improves both the precision and the bias where ACE error and the standard deviation both remain within 2% and 6%, respectively. The ACE estimation using PredictUS window outperforms the estimation from the small window and achieves performance equivalent to that obtained from the large window.

In case II, the difference between the ACE measures obtained from small window (5λ) and the large window (10λ) is more prominent, where small window results in a bias up to 7.3% and standard deviation as large as 19.7%. Interestingly, the PredictUS window, using original data of length 5λ only, can attain a precision within 5.5%, slightly larger than that obtained from the large window (4%). The PredictUS window gives moderate bias (within 6%) for ACE < 1.1 dB/cm/MHz.

Finally, for case III, the PredictUS window only includes 2.5λ original data, however, achieves a standard deviation within 10%. Although, the increasingly accumulated error in the prediction of RF data affects the accuracy, resulting in a larger ACE error compared to the small original window.

In summary, compared to the optimum trade-off, the PredictUS method can reliably maintain similar precision and accuracy while improving the resolution to 75% of the optimum value. Moreover, the proposed method can still achieve comparable precision and accuracy for low ACE values with a resolution improvement of 50%. However, reducing the resolution to 25% of the optimum exhibit degraded ACE measurement, which can be attributed to the error accumulation in the n-step ahead prediction. Unlike the case of RF data prediction, high ACE does not have any distinct effect on the performance of ACE computation.

4   Conclusion

We propose a novel QUS parameter estimation method, PredictUS, utilizing ultrasound RF signal prediction. The method shows promising results by predicting larger RF windows from the smaller ones. We conduct a proof-of-concept study based on extensive simulation analysis. The proposed sequence-to-sequence convolutional neural network based on WaveNet model was able to estimate RF signal samples to a reasonable accuracy and therefore improve the power spectral estimate. A resolution reduction, as high as 50%, while maintaining comparable estimation precision introduces a paradigm shift by challenging the insistent trade-off between precision and resolution, inherent in ultrasound spectral estimation. Future research will address the issue of error accumulation in the n- step prediction by further improving the CNN structure.

References

1. Oelze, M.L., Mamou, J.: Review of quantitative ultrasound: envelope statistics and backscatter coefficient imaging and contributions to diagnostic ultrasound. IEEE Trans. Ultrason. Ferroelectr. Freq. Control **63**(2), 336–351 (2016)
2. Deeba, F., et al.: Attenuation coefficient estimation of normal placentas. Ultrasound Med. Biol. **45**(5), 1081–1093 (2019)
3. Deeba, F., et al.: SWTV-ACE: spatially weighted regularization based attenuation coefficient estimation method for hepatic steatosis detection. In: International Conference on Medical Image Computing and Computer-Assisted Intervention (2019)
4. Yao, L.X., et al.: Backscatter coefficient measurements using a reference phantom to extract depth-dependent instrumentation factors. Ultrason. Imaging **12**(1), 58–70 (1990)
5. Oelze, M.L., O'Brien Jr., W.D.O.: Defining optimal axial and lateral resolution for estimating scatterer properties from volumes using ultrasound backscatter. J. Acoust. Soc. Am. **115**(6), 3226–3234 (2004)

6. Liu, W., Zagzebski, J.A.: Trade-offs in data acquisition and processing parameters for backscatter and scatterer size estimations. IEEE Trans. Ultrason. Ferroelectr. Freq. Control **57**(2), 340–352 (2010)

7. Wear, K.A., Wagner, R.F.: A comparison of autoregressive spectral estimation algorithms and order determination methods in ultrasonic tissue characterization. IEEE Trans. Ultrason. Ferroelectr. Freq. Control **42**(4), 709–716 (1995)

8. Oord, A.V.D., et al.: WaveNet: a generative model for raw audio. arXiv preprint arXiv: 1609.03499 (2016)

9. Borovykh, A., Bohte, S., Oosterlee, C.W.: Conditional time series forecasting with convolutional neural networks. arXiv preprint arXiv:1703.04691 (2017)

10. Treeby, B.E., et al.: Modeling nonlinear ultrasound propagation in heterogeneous media with power law absorption using a k-space pseudospectral method. J. Acoust. Soc. Amer. **131**(6), 4324–4336 (2012). Medical physics of CT and ultrasound: Tissue imaging and characterization

11. Nam, K., Zagzebski, J.A., Hall, T.J.: Simultaneous backscatter and attenuation estimation using a least squares method with constraints. Ultrasound Med. Biol. **37**(12), 2096–2104 (2011)

Author Index

Printed in the United States
By Bookmasters